$\frac{51}{320} = 211$

$= \frac{46}{211} = .22 \sec$

$\frac{?-a^2)}{16 - 025}$

$.14 = .42 \text{ abbs.}$

$\frac{400}{42} = 1000 \text{ amp}$

length of wire $= 6$ cm

$\frac{1}{2}$ H on out

conductors

46 cm

$\therefore V = A \log_e \frac{a}{b}$

diff..... velocity

for potential gra...

time taken to ...

$d\tau = \frac{dr}{u \cdot \frac{??}{?\,?}}$

$\therefore \int_0^t dt = \int_a \frac{r\,dr}{u\,v\,?}$

$\therefore t = \frac{1}{2u \cdot v \cdot ??}$

$= \frac{\log_e \frac{a}{b}}{\pi u \, r}$

say approx cleared of

4 cm from

$B = $ ...

Breadth of tube $= .4$ c

...ions travel .4 cm while

Like Mark Oliphant, Andrew Ramsey was born in South Australia, and he currently lives beside a school that the young Oliphant attended in Adelaide. He has been a journalist for thirty years, based in Adelaide and Melbourne, and has had his articles published in major national newspapers in Australia, the United Kingdom and India. He has also worked as a political speechwriter and in corporate communications roles in universities. He is the author of *The Wrong Line* and *Under the Southern Cross – The Heroics and Heartbreak of the Ashes in Australia*.

THE BASIS OF EVERYTHING

Rutherford, Oliphant and the Coming of the Atomic Bomb

ANDREW RAMSEY

HarperCollins*Publishers*

HarperCollins*Publishers*

First published in Australia in 2019
by HarperCollins*Publishers* Australia Pty Limited
ABN 36 009 913 517
harpercollins.com.au

Copyright © Andrew Ramsey 2019

The right of Andrew Ramsey to be identified as the author of this work has been asserted by him in accordance with the *Copyright Amendment (Moral Rights) Act 2000*.

This work is copyright. Apart from any use as permitted under the *Copyright Act 1968*, no part may be reproduced, copied, scanned, stored in a retrieval system, recorded, or transmitted, in any form or by any means, without the prior written permission of the publisher.

HarperCollins*Publishers*
Level 13, 201 Elizabeth Street, Sydney NSW 2000, Australia
Unit D1, 63 Apollo Drive, Rosedale, Auckland 0632, New Zealand
A 53, Sector 57, Noida, UP, India
1 London Bridge Street, London, SE1 9GF, United Kingdom
Bay Adelaide Centre, East Tower, 22 Adelaide Street West, 41st floor, Toronto, Ontario, M5H 4E3, Canada
195 Broadway, New York NY 10007, USA

A catalogue record for this book is available from the National Library of Australia.

ISBN 978 1 4607 5523 5 (paperback)
ISBN 978 1 4607 0955 9 (ebook)
ISBN 978 1 4607 8093 3 (audio book)

Cover design by Mark Campbell, HarperCollins Design Studio
Front cover image: Cavendish Laboratory staff, Cambridge, 1932, with Ernest Rutherford highlighted in front row and Mark Oliphant in second row, courtesy Oliphant Papers, Rare Books and Special Collections, University of Adelaide Library, Australia
Back cover image: Mushroom cloud of atomic bomb dropped on Nagasaki, Japan, 9 August 1945, from shutterstock.com
Endpapers: Pages from Ernest Rutherford's Cavendish Laboratory notebooks, 1896, courtesy Oliphant Papers, Rare Books and Special Collections, University of Adelaide Library, Australia
Typeset in Adobe Garamond Pro by Kirby Jones
Printed and bound in Australia by McPhersons Printing Group
The papers used by HarperCollins in the manufacture of this book are a natural, recyclable product made from wood grown in sustainable plantation forests. The fibre source and manufacturing processes meet recognised international environmental standards, and carry certification.

When we have found how the nucleus of atoms is built up we shall have found the greatest secret of all ... We shall have found the basis of everything – of the earth we walk on, of the air we breathe, of the sunshine, of our physical body itself, of everything in the world.

ERNEST RUTHERFORD

To Beverley
For quietly inspiring, and forever encouraging

CONTENTS

Prologue ... 1

1. Colonial Boys ... 11
2. The World Awaits ... 26
3. 'Rabbit from the Antipodes' ... 45
4. 'They'll Have Our Heads Off' ... 58
5. The Atom Smasher ... 72
6. A Benevolent Lord ... 90
7. 'A Rare Quality of Mind' ... 103
8. String and Sealing Wax ... 112
9. A Meeting of Minds ... 129
10. The Golden Year ... 147
11. Fusion ... 158
12. Tyranny's Dark Clouds ... 170
13. The Crown Begins to Slip ... 177
14. 'Requiem Aeternam' ... 197
15. 'A Show of My Own' ... 206
16. The Decisive Difference ... 221
17. 'Shouldn't Someone Know About This?' ... 239
18. MAUD ... 256
19. 'Meddling Foreigner' ... 277
20. A Misguided Mission ... 291
21. Manhattan ... 303
22. 'Death, the Shatterer of Worlds' ... 323
23. 'We Have Killed a Beautiful Subject' ... 329

Epilogue ... 346
Endnotes ... 354
Bibliography ... 367
Acknowledgments ... 373

PROLOGUE

South Australia, 1925

For the third time in as many decades, Ernest Rutherford sailed into the port of Adelaide while being buffeted by the gathering headwinds of imminent change. He was certainly grateful to meet land after a torrid Indian Ocean crossing. So tempestuous was the voyage from Cape Town aboard the SS *Ascanius* that a section of the ship's teak railing was ripped from the deck during a south-westerly gale. It meant the physicist and his wife, Mary, had spent much of the Australia-bound leg bunkered down in their first-class cabin.

Rutherford therefore inhaled deeply on the brackish breeze that fluttered almost apologetically across Port Adelaide's Outer Harbor as he unsteadily descended the gangplank slung from the steamer. It had not long gone nine o'clock on the morning of 3 September 1925, and he paused momentarily to ponder the changes – to the crowded passenger wharf directly in front of him; to that portion of an extraordinary life story behind him; to the mind-bending details he had unearthed about the very world around him – wrought during the thirty years since his initial visit.

Adelaide had been his first port of call after leaving New Zealand in 1895 as an unknown and uncertain science student bound for a brash adventure at Cambridge University's Cavendish Laboratory, where he had earned instant notoriety as the first scholar from such distant dominions to be admitted to the exclusive institution's hallowed cloisters. When Rutherford had returned to Adelaide in 1914, he had come cloaked in the acclaim rightly afforded the first human to prove that an atom of matter could be divided. That accomplishment had earned the son of a peripatetic flax farmer a knighthood in his adopted English homeland, and a Nobel Prize for his pioneering exploration of the sub-atomic world. But on that occasion, when he had arrived as figurehead of a British scientific delegation, he had carried foreboding as well as fame. The political unrest boiling in Europe upon his departure that northern summer had exploded into conflict while he was at sea. It was a conflagration that would rapidly escalate into the war supposed to end all wars.

So Rutherford, who had turned fifty-four just days earlier, might have justifiably wondered what consequences would flow from his third sojourn in Adelaide in 1925, the launching point for his series of lectures to be given across Australia and in his homeland, New Zealand. Those addresses would shed light on the astounding experimental work he was overseeing in his now-fabled Cavendish Laboratory. And it was those findings that had already fundamentally changed the world's understanding of, and relationship to, the universe's building blocks. Rutherford's genius had revealed to science, to everyone, the mysteries of atoms and their constituent parts. It was information that had hitherto remained unseen and unknown by humankind, and with it had come whispered warnings about the huge power these minuscule particles held and the danger that might be awoken if it were let loose.

Yet for all his insights, Ernest Rutherford did not foresee the chain reaction that would be set in motion by his Adelaide stopover. Indeed, the footfall from that early spring morning that would ultimately echo through history was muffled amid the back-slapping reverence of the local dignitaries who comprised his quayside welcoming party. Barely had Rutherford alighted upon dry land when he was ushered to a car and driven to a formal civic reception at Adelaide's grand Town Hall.

While his wife made a head start on the couple's planned reunion with their respective families in New Zealand that would follow the lecture tour, Rutherford was chaperoned to the mayoral gathering. Adelaide was in a celebratory mood – though not altogether in expectation of the two lectures the famous visitor was scheduled to deliver. Rather, spirits were lifted by the bud-burst of pure white almond blossoms across the suburbs and throughout their hills backdrop, confirming not only the passing of winter, but also the imminent opening of the annual Royal Adelaide Agricultural and Horticultural Show.

The Town Hall event was attended by the most eminent figures from the city's sole university, among them Antarctic pioneer Sir Douglas Mawson, then Professor of Geology and president of South Australia's Royal Society. Mawson's discoveries upon earth's last, vast unexplored continent more than a decade earlier afforded him significant renown. But now a new generation of intrepid explorers, led by Rutherford at the Cavendish, was gaining similar fame by probing sub-atomic territory that could not be seen, much less traversed.

Adelaide's Lord Mayor, Charles Glover, could not resist the opportunity to indulge in a little cultural appropriation by decreeing that Australians might 'advance some sort of claim to nationhood' over Rutherford through his New Zealand pedigree. Mawson, in turn, showered praise upon the slightly self-conscious guest of honour, who responded to the rapturous applause by describing himself 'as a comparatively insignificant unit – an atom of the universe as I am today'.[1]

Mawson would have none of it. 'No-one [is] more distinguished in the realm of science today than Sir Ernest Rutherford,' he declared. 'In fact, I doubt there has ever been anyone more notable – he is so fundamental, so thorough and complete that his work will stand for all time.'[2]

But if Rutherford needed confirmation of just how fleeting eminence can be, it came when the reception concluded and he waited beneath the Town Hall's heavy stone portico for a taxi to the South Australian Hotel, where a bundle of correspondence sent on from Cambridge awaited. As a local newspaper later revealed, 'It was about one o'clock and hundreds of people were hurrying to lunch, unaware that they were passing one of the world's most distinguished scientists. "If he had only been Jack Johnson, the pugilist, or a cricketer, it would require a posse of police to clear the footpath", exclaimed an attendant at the reception to his companion.'[3]

Accompanying Rutherford as local liaison agent was Kerr Grant, Adelaide University's Elder Professor of Physics. Grant well knew of

Rutherford's hectic Australian itinerary, yet he also understood how rarely such an esteemed figure set foot in his domain. Seizing his chance, the professor asked Rutherford if he would consider making an informal visit to his physics department the following day – Friday, 4 September. To his delight and surprise, Rutherford agreed without hesitation.

This unexpected, unscripted engagement would reshape the future.

* * *

Such was Rutherford's celebrity at that time that details of his landing, his welcome reception, and his inaugural lecture were recounted in studied detail by the next morning's newspapers. Mark Oliphant found those accounts so engrossing that even the rattle and roll of his daily steam-train commute to Adelaide University was not sufficiently severe to prise his eyes from the newsprint. Oliphant habitually spent the half-hour journey from his new marital home at Glenelg thoroughly immersed in the latest international science journals, or studying textbooks that informed his work as a researcher and demonstrator in the university's physics department. On this Friday morning, however, as spring dawned and the city readied for its showpiece annual carnival, the twenty-three-year-old's attention was reserved exclusively for the story on Ernest Rutherford.

While his fellow travellers buzzed with chatter about the next day's round of football matches, or craned for glimpses of the recently finished fairgrounds where the Royal Adelaide Show would come alive over the weekend, Oliphant's interest was strictly academic. And the topic that captivated him was the man he knew well from Rutherford's myriad published works, but whose presence he never envisaged he might one day share.

On having reached his destination at Adelaide's railway terminus and climbed down from his carriage, Mark Oliphant neatly folded the pages of newspaper into perfectly proportioned rectangles and tucked them snugly beneath the arm of his tweed jacket. As he struck out on the gentle, ten-minute walk to the university, he carried no inkling that that 'one day' had dawned. And that his life would soon be irrevocably changed.

PROLOGUE

* * *

That very morning, Rutherford paid his promised visit to the physics department, then located on the ground floor of what is now Adelaide University's Mitchell Building. Kerr Grant proudly detailed work being undertaken at the new biochemical laboratory, and showed off the dedicated physics and engineering building that was nearing completion a short walk from the current, confined physics quarters. Some students and laboratory staff who were tinkering with purpose-blown glassware and antique brass calibration tools immersed themselves in their tasks so as not to stare at the famous visitor wandering among them. Others, doubtless startled by Rutherford's thunderous voice and booming laugh, which bounced off the worn wooden floorboards and bare stone walls, attached themselves to Kerr Grant's tour group. The party numbered around 200 by the time it reached the new building, where Rutherford launched into an informal address that featured animated updates on the work being carried out by his 'boys' at the Cavendish Laboratory.

When Rutherford spoke, the excited hubbub that had built behind him during his inspection of premises old and new was stilled to silence. Even the artisans adding the finishing touches to the pristine experimental space downed tools as the famed physicist explained the latest sub-atomic secrets and how they had been revealed. He detailed how particles previously unknown to humankind were being violently collided at the Cavendish, and the resultant pieces forensically examined. The stories that Rutherford told carried his hushed audience from credulity to unshakeable belief, such was the passion and authority with which he spoke.

There was no-one, among the enthralled audience, more moved by Rutherford's presence and propositions than the young man who stood silently on the group's perimeter, a tall, looming presence, with a vertiginous plume of rust-coloured hair. His eyes, framed behind gold-rimmed spectacles, sparkled at each of Rutherford's wholehearted jokes, and grew wider with every revelation of the boundaries being pushed at Cambridge.

Like that day's eminent guest, Mark Oliphant hailed from humble origins with no ancestral connection to higher learning, and had overcome numerous obstacles to win a place at university. In the almost three years since graduating under Kerr Grant's guidance with a first-class honours degree, he had held positions as physics department researcher and demonstrator. But as he stood impassively among the freshly installed workbenches and few pieces of scattered apparatus that Friday afternoon, Mark Oliphant was rendered as awestruck as any first-year novice.

Amid the buttoned-down sensibilities of Adelaide University in the 1920s, where students – save for the sizeable female cohort – routinely sported suits and ties, Mark Oliphant showed a free spirit. While the young man's skill at crafting equipment and conducting experiments had won the attention of senior staff within the small physics faculty, his manner had distinctly polarised his peers. From the outset, some of his fellow students had regarded him as a high-spirited prankster. He had proven himself adept at hurling water bombs and firing 'grape guns' – narrow glass tubes from which the fruit could be forcibly expelled, like a blunted blow dart. In those varsity play fights, he was also known to discharge the foaming contents of a fire extinguisher on his enemies, if such a weapon was conveniently at hand. During his early student days, he had once leaped from a low-rise rooftop with only a home-made parachute to arrest his descent.

Others described Oliphant as aloof and prone to flashes of temper that would quickly subside, one fellow student claiming: 'In the laboratory he was often bossy, abrasive and rude. He used to give orders, I thought him over confident, yet he could also be a pleasant enough chap.'[4] Another, Walter Schneider, described him as a 'big, burly, almost bear-like boy – a pleasant companion, but a loner'.[5] Predominantly, Oliphant was seen by his peers as a larger-than-life figure. What was also obvious in the independently minded student was his preparedness to challenge accepted wisdom.

Yet when Sir Ernest Rutherford carried out his tour of the physics department, Mark Oliphant was rendered a mute onlooker. As Oliphant would later recall of that epochal meeting, 'at that time, members of the University as humble as I, were not introduced to such illustrious visitors'.[6]

It was Rutherford's character and conviction more than any polished presentation or rhetorical flourishes that won over Oliphant that portentous day. Indeed, during the years to follow, Oliphant learned that Rutherford's presentations were often as likely to infuriate as illuminate. What made them resonate so clearly with those who understood his vision, however, was the unashamed love he felt for his sub-atomic subject matter.

'Rutherford was not a good lecturer,' Oliphant would reflect. 'He hummed and he erred and he haa-ed, and he stood back from the blackboard and looked at it for periods of time, but there was something about him that made you feel that he cared. That this subject of physics generally, of understanding nature was something that really mattered to him. It was his life. It was inspiring in that sense.'[7]

That inspiration was indelibly transmuted to Mark Oliphant during that brief encounter. As Ernest Rutherford sketched a vivid word picture of his Cavendish Laboratory and the history being written within it, Oliphant found himself deeply drawn to the world of atomic investigation – a field for which Cambridge University had become internationally famous, but in which Oliphant held little direct experience at that point. He knew enough, however, to understand immediately where his destiny resided.

'It was so inspiring to hear a man of that calibre talk when I'd been taught by ... ordinary sort of professors who were quite good in their way but not Presidents of the Royal Society ... the man who had unravelled the whole story of the way in which uranium changed, over time, into lead,' Oliphant recounted years later. 'He was so generous in giving his praise to the people who were his students, who worked with him and [were] doing the jobs, that I thought to myself then, "That's the man I want to work with." So from then on, my efforts were directed towards trying to get a scholarship to go to Cambridge.'[8]

Even though they did not exchange a word during Rutherford's short visit, Oliphant intuitively felt their futures would become entwined. He sensed that the great physicist was someone from whom he could learn not only about science, but about the world and its vast complexities. As Oliphant would later confirm: 'He became a hero to me and, later on, much more than a hero. He was a man that I grew really to love ... he was so inspiring. Wherever he was, he always was the dominant figure. Not in any sort of domineering way, but simply through sheer personality.'[9]

* * *

Within three years, Ernest Rutherford and Mark Oliphant would forge a partnership that transcended that of master and pupil, and, over the subsequent decade, grew more akin to that between father and son. These two men, born thirty years and 3000 kilometres apart – Rutherford on the South Island of New Zealand, Oliphant in the similarly progressive settlement of South Australia – shared more than a passion for science and a drive to unlock the deepest mysteries of nature. They were also shaped by the mutual experiences of colonial settlement, modest upbringing, rudimentary education and unfettered curiosity for their respective worlds.

There was nobody whom Ernest Rutherford came to trust more implicitly, or rely upon more completely, than Mark Oliphant. In turn, there was no-one Oliphant more unashamedly revered, or whose approbation he sought more than that of Rutherford. It was Rutherford who ignited Oliphant's research passion when happenstance brought the pair together at Adelaide University. It was Oliphant's vision and pragmatism that immediately endeared him to Rutherford upon their first meeting in the sanctum of Cambridge's Cavendish Laboratory. It was Rutherford who ushered Oliphant into the hallowed ranks of nuclear physicists, and who earmarked him to continue his peerless legacy. And it was to Oliphant that Rutherford turned in his private moments of self-doubt and personal grief.

Their friendship was fundamental to the last true golden age of institutional science, when the race to crack the secrets of nuclear physics and harness their frightening force gripped the globe. That quest, of which Rutherford's Cavendish Laboratory was the core, would ultimately yield breakthroughs unimaginable a century ago – television, computers, smartphones, wireless technology, satellite tracking, cancer treatments, medical imaging and the internet among them. It also produced an arsenal of weaponry that bears the two men's distinct fingerprints, and which continues to cast a malevolent shadow over the planet's very future.

Rutherford and Oliphant's story is that of the nuclear age's conception and birth. Yet its roots took hold in a time of similarly bold ideas and unbending spirit – an age when Britain's utopian nineteenth-century vision was transplanted into its furthest colonies.

1

COLONIAL BOYS

South Australia and New Zealand, 1871 to 1916

Edward Gibbon Wakefield was serving three years in Newgate Prison for the abduction of a fifteen-year-old heiress – whom he had then secretly married in the hope of securing her inheritance – when he devised the radical colonisation model that gave birth to South Australia. This was something of a paradox, given that a central pillar of 'systemised colonisation' was the explicit exclusion of convicts.

What also set the proposed new settlement apart from Australia's existing colonies was the novel idea of preselling parcels of surveyed land, to attract immigrants of financial means. Capital raised from those sales would subsidise the relocation of less wealthy families, thereby providing South Australia with a willing labour force.

While the penal colonies had been devised to ease pressure on Britain's overstocked prison system, the systemised settlements were designed to solve a similar crisis in its choked industrial cities. 'Here we are, three or four in a bed,' Captain William Gowan told the inaugural public meeting of the South Australia Association in late June 1834. 'We cannot walk along the streets, but we are jostled by some person thrusting his elbow into our side.'[1]

That urgency was reflected in the speed at which Wakefield's idea evolved from concept in the late 1820s to colony in 1836. The first shipload of South Australia's free settlers dropped anchor at Holdfast

Bay a few days after Christmas of that year – at which time little of the required survey work had begun. This, coupled with a low asking price of £1 per acre (less than AU$200 today), meant the venture risked collapse under a weight of debt before it had even found its feet. It also meant that Wakefield, the scheme's architect, distanced himself from further involvement and shifted his sights to nearby New Zealand.

Not six months after the hopeful South Australian colonists had celebrated their arrival with a drunken ship's party on the mosquito-plagued shores of Holdfast Bay, Wakefield oversaw the establishment of the New Zealand Association. His enthusiasm for the new endeavour led him to proclaim that the volcanic islands on the fringe of the South Pacific Ocean represented 'the fittest country in the world for colonisation'.[2]

Following four years of complex wrangling, a promising location named in honour of Vice-Admiral Horatio Lord Nelson was planned on New Zealand's South Island. Come the dawn of 1842, an advance colonisation party had built more than 100 basic huts at Nelson, and within months the community numbered around 500 immigrants.

As the frontier township grew, there came a need for qualified artisans to provide its basic infrastructure. The New Zealand Association placed advertisements in newspapers throughout Britain, and it was a call for a general-purpose wright – proficient in shaping timber, iron and steel – that caught the attention of George Rutherford in the overcrowded Scottish city of Dundee. With his wife, Barbara, and their four children, he boarded the 470-ton *Phoebe*, which sailed for Nelson on 16 November 1842. They arrived in port on a warm, clear autumn afternoon in late March of the following year.

No sooner had the family made footfall in Nelson than they began an onward journey of almost fifty kilometres across Tasman Bay to the fertile soils and soaring forests around Motueka. It was there that they settled in a mud and raupo (swamp-reed) hut, and that George went to work on the sawmill he had been hired to help construct.

Despite a decade of gruelling labour that included bouts of desperate poverty, George eventually saved sufficient money to purchase twenty acres of land at Waimea South, near the Wai-iti River, in late 1854. The property faced the main track leading to Nelson, twenty-five kilometres away: a thoroughfare since named Lord Rutherford Road.

South Australia, 1840s to 1850s

Across the Tasman Sea, the fragile colony of South Australia continued to be buffeted by economic and social turbulence. The shortfall in the land sales needed to bolster its barren coffers was soon compounded by a dramatic population exodus to the lucrative goldfields of Victoria and New South Wales.

The resultant need to replenish a rapidly diminishing workforce saw assisted immigration grow from around 4600 in 1853 to almost double the following year. Among that 1854 intake was James Smith Olifent, a grocer by trade from the Dover region of county Kent, who, accompanied by his wife, Eliza, and their children travelled aboard the three-masted barque *Ruby* and disembarked at Adelaide's port in late March.

James Olifent abandoned the grocery business and found work at Adelaide's destitute asylum, newly established to tackle the colony's escalating social dysfunction. James would eventually become superintendent of the grim institution, housed in a complex of austere, double-storey buildings across the road from the governor's grandiose quarters, where his great-grandson would be installed as vice-regal resident more than a century later.

Within two years of the Olifents' arrival, a government-funded library opened next door to the asylum, in the South Australian Institute. Soon afterwards, as pledged by the colony's founders, a circulating library was also established. Collections of books, magazines and newspapers were conveyed to country towns and outlying settlements in an innovative scheme believed to be a global first. By 1866, there were twenty book boxes circulating throughout South Australia; seven years later, that number had increased fourfold.[3]

Also among South Australia's first statutory authorities was the Central Board of Education. Annual enrolments at accredited schools stood around 3300 shortly before the Olifent family's arrival, but had almost tripled within a decade.

The term 'book colonies', sometimes sneeringly applied to South Australia and the New Zealand communities settled under the systemised scheme, stemmed from their contrived genesis by ink on paper rather than boot leather across unexplored ground. However, it could also be taken as reference to the ready availability of schooling and reading material in those ambitious settlements, which sought to change the manner in which Britain expanded its empire.

Certainly, books and learning were integral to shaping the brilliant minds of Ernest Rutherford and Mark Oliphant. While both young men were encouraged by supportive parents and driven by innate curiosity, the opportunities afforded them by public schooling – and in Oliphant's case, the crucial influence of a public library – suggested the name 'book colonies' might also reflect these settlements' nobler aspirations.

New Zealand, 1865 to 1882

The school that served the farming community of Waimea South, where George Rutherford had built a timber home for Barbara and their brood, which had increased to five boys and three girls, was sited in the neighbouring hamlet of Spring Grove. Among those who taught the school's 100 or so students were a widow, Caroline Thompson, and her twenty-year-old daughter Martha. When Spring Grove's headmaster died suddenly in 1865, Martha Thompson was briefly appointed as sole senior teacher. However, she resigned after two months in the prestigious role due to her forthcoming marriage to George and Barbara Rutherford's third son, James.

The home that James and Martha then built on a portion of George Rutherford's twenty-acre allotment provided sleeping space for the first eight of the couple's eventual twelve children. The fourth of those, registered through clerical oversight as 'Earnest', was born at the house on 30 August 1871.

South Australia, late 1860s to 1901

By the late 1860s, Adelaide's General Post Office had become the hub of telecommunications between Australia and the wider world. Before the spread of wires and poles and the roll-out of under-sea cables, messages from the mother country were carried by mail steamers and offloaded at King George Sound (now Albany). From there, they would be couriered to Adelaide for faster dissemination to the eastern seaboard.

This costly and time-consuming system required one of South Australian Postmaster-General Charles Todd's employees to regularly make the 5000-kilometre round trip to and from the mail steamer's terminus. One of those clerks was Harry Smith Olifent, son of James and Eliza, who took the job immediately after leaving school. Rapid advances in ocean transport and telecommunications rendered the role redundant by the 1870s, and Harry Olifent was reassigned to a desk job at Todd's GPO.

In addition to serving as an office bearer for the Freemasons' mother lodge, Harry Olifent was an avid member of a literary society that met at a Congregational church in central Adelaide. It was through this social circle, one of many that drew together 'book colony' folks with like-minded values, that Harry met his future wife, Alice Robinson. Reading and learning became highly prized in the couple's single-storey cottage in the inner eastern suburb of Dulwich, where they raised seven children.

Despite his modest GPO earnings, Harry was somehow able to enrol their second son, Harold George Olifent, at Prince Alfred College, one of Adelaide's most prestigious private schools. Young Harold, who later became known to all as 'Baron' in recognition of his tall stature and dignified demeanour, clearly showed sufficient academic potential for his parents to make major financial sacrifices in order to nurture it.

Yet, to his enduring frustration, Baron would find himself a lifelong civil servant. His only dalliance with manual work was a fleeting, impulsive dash to Coolgardie in the West Australian goldfields during the rush of the 1890s. He returned not with a fortune, but with a small, solitary nugget, which he later bequeathed to his first-born son.

Around the turn of the twentieth century, when Baron is believed to have adopted the alternative spelling 'Oliphant', such an administrative matter did not require formal lodgment with authorities. The name 'Olifent' was recorded on the marriage certificate for Harold George (Baron) and Beatrice Tucker when they were wed on 27 December 1900, just days before Australia celebrated Federation.

The couple had met when Baron, like his father, joined a literary society, operated by St Paul's Church in central Adelaide. The church adhered to the Anglo-Catholicism into which the Olifent children had been raised, and of which Baron had become a devout follower. Beatrice was the daughter of a headmaster at a government-run school and she, not unlike Ernest Rutherford's mother, Martha, began her working life as sole teacher in an isolated community – at Hawker, a staging post almost 400 kilometres north of Adelaide. Her commitment to the church did not run as deep as Baron's, but they shared similar reading tastes, as well as views on politics and social issues very much in step with their respective educated, working-class upbringings.

'Oliphant' is the surname registered for the couple's first child, born on 8 October 1901. The boy was named Marcus Lawrence Elwin Oliphant, in honour of author Marcus Clarke, whose sprawling novel *For the Term of His Natural Life* was the era's definitive account of the brutal penal system founded through British imperialism on the far side of the globe. The couple's academic ambitions for their son were apparent from his birth.

New Zealand, 1876 to 1882

The rapid spread of railways into New Zealand's countryside in the decades after Nelson's settlement meant lucrative bridge construction work for James Rutherford. Therefore, as his family moved deeper into the South Island's heart, Ernest Rutherford began school, aged five, at Foxhill, fifteen kilometres from Spring Grove.

The forty or fewer students who knew Ernest Rutherford at the single-room Foxhill Primary School later recalled a most unexceptional boy. Quiet and self-contained, young Ernest rarely engaged in sports or schoolyard games, and preferred to immerse himself in a book while sitting beneath a tree.

Among his treasured primary school possessions was a well-thumbed primer simply entitled *Physics*, written by Scottish scientist and educator Balfour Stewart. As her son's eminence grew through adulthood, Martha

Rutherford would cherish the boyhood keepsake, which bore the simple inscription 'Ernest Rutherford, July 1882' inside its leather cover.

As a boy, Ernest Rutherford's propensity to dribble when lost in distant thoughts, coupled with his solitary nature, saw him nicknamed 'Dopey'.[4] He did, however, boast an advantage over many of his student peers, in that his mother had taught her children reading, writing and basic arithmetic before any of them first walked through a school gate.

Come evening at the family home, Martha Rutherford would gather her progeny around the household hearth, and challenge them with knowledge quizzes and spelling bees. As they progressively reached school age, she would invigilate as they completed their homework and ensure the afternoon and weekend farm chores did not take precedence. On Sundays, when the strictly observed Sabbath meant no frivolity in the Anglican household, Martha would lead the family singalong around her prized Broadwood piano, imported from England and bought with money she had saved during her teaching tenure at Spring Grove.

James Rutherford actively encouraged his wife's impromptu tutorials, conspicuously aware that his own family's semi-nomadic lifestyle had denied him a formal education. As a consequence, by the school leaving age of eleven, he had been competent at reading but unable to write. The reality that he was therefore destined to remain a tradesman meant that James Rutherford's family also had to accept his restless quest for reliable income.

In 1881, James took a ten-year lease on flax-growing land near Havelock, 100 kilometres away via a treacherous dirt track. He left Martha and their eleven children at Foxhill while his new enterprise took shape, and was only able to visit them every three months or so. The Rutherford children were thus almost exclusively in their mother's care.

After two years of separation, James arranged for his wife and children to sail from Nelson to Havelock aboard a small coastal steamer. Ernest was nearing his twelfth birthday in mid-1882 when he began at Havelock's school, which was almost twice the size of Foxhill's. As he once more settled into new surrounds, his detached manner and quiet nature landed him the nickname 'Windy'.

South Australia, 1912 to 1916

Mark Oliphant was also on the brink of adolescence when he entered a single-teacher rural school in 1912. His family had relocated from Adelaide's inner suburbs to Mylor, an experimental hills community twenty-five kilometres from the city centre. The succession of modest rental homes that Baron and Beatrice Oliphant had occupied in Adelaide's inner southern suburbs could no longer accommodate their five sons in any pretence of comfort.

Baron's job as a ledger clerk with the city's water department meant he was unable to make the hours-long commute from the hills each day, so while Beatrice and the boys spent their weekdays in bucolic isolation, he resided with his parents at Dulwich. During those years, the family would only be united during weekends and annual holidays.

Before moving to the hills, Mark had learned he was congenitally and completely deaf in his left ear. Compounding his challenges was the astigmatism diagnosed at Mylor, a condition that blurred his vision. His short-sightedness also prevented him from reading the blackboard from the classroom's back rows, to which taller pupils were assigned. Oliphant's need for wire-rimmed glasses meant he was dubbed 'Four Eyes' in the schoolyard and that, like Rutherford at the same age, he was little interested in sports. It was the adulteration of his surname, rather than any slight on his unathletic disposition, that led to his other nickname of 'Roly Poly'.

While Rutherford would develop a fondness for rugby when he reached senior school, Oliphant's flirtation with Australia's native football code was brief, and almost painfully comical. In one of his few Australian Rules football matches, his poor eyesight led him to kick the ball to the advantage of the opposing side, which prompted one of his aggrieved teammates to threaten to punch him.

With Baron an absentee father for much of the family's four years at Mylor, Beatrice Oliphant took on the role of single parent, just as Martha Rutherford had done in similar circumstances. Beatrice's schoolteacher training meant that the Oliphant boys, like the Rutherford children decades earlier, developed an appreciation for learning and literature.

Where Martha Rutherford had delighted in reminding her offspring that 'All knowledge is power', Beatrice Oliphant's refrain when her sons posed a question was 'Well, let's look it up'. The imperative to source credible, factual detail in order to solve any problem became a cornerstone of all Mark's future endeavours.

Beatrice became the stable domestic presence in her eldest son's life. 'My mother had far more influence on me in reality than my father,' Mark would recall. 'I'd been brought up in a family where the mother was a very important person indeed, and for us children … much more important than our father.'[5]

The other influential figure during his time at Mylor Primary School was the head teacher, Bernard McCaffrey. When in 1973, as Governor of South Australia, Mark returned to his former school to open its new library, he lavished praise on McCaffrey, who had introduced him to the words of Coleridge, Stevenson and Tennyson, and nurtured within him a deeper love of literature than that already instilled by his parents. The ruddy-faced Irishman taught his pupil 'to distinguish the good from the bad, the works of genius from the mediocre. It was here in Mylor that books became not only my companions, but my very life.'[6]

New Zealand, 1883 to 1886

Similar influences were wielded upon Ernest Rutherford at a comparable age. On landing at Havelock in 1883, the Rutherford children found themselves under the tutelage of the school's sole teacher, Jacob Reynolds, aged just twenty-seven. With James Rutherford's work at the remote flax mill requiring his presence on site for six days of every week, Ernest developed a close rapport with his charismatic young schoolmaster.

The singular focus that had characterised Ernest Rutherford's childhood was channelled into organised study, as the twelve-year-old student began to understand that academic achievement offered him the surest path away from his dad's life of relentless physical toil and economic uncertainty. What also became apparent at this time was Ernest's capacity to block out all distractions and narrow his concentration to the single task in front of him. This was likely honed in a household shared with so many siblings, and then at Havelock school, which was so oversubscribed that its senior students took their lessons in the town hall.

The previously self-absorbed boy responded to the tough academic benchmarks set by Reynolds, and the methods his teacher employed to help him reach them. Reynolds's unapologetic quest for excellence saw him criticised by some indignant Havelock parents for demanding that students complete homework during their summer holidays. Yet despite being two years younger than others who sat the standard six examinations to provide the qualification, if not the means, to progress to secondary school, Ernest passed with quiet certainty.

South Australia, 1910s

At around the same age that Ernest Rutherford was nearing secondary school, Mark Oliphant's horizons were also broadening. The fascination for the world, and curiosity as to his place within it, that grew as he approached teenage-hood had been nurtured by what was in some ways an unconventional upbringing.

As Baron Oliphant sought escapism from his stifling desk job, his political views had become increasingly shaped by the Fabians' social-democratic ideologies. His search for further answers had taken him beyond the High Anglican Church and into the faddishly popular Theosophy movement. For years, Baron served as president of Adelaide's Theosophical Society, which promoted the revival of ancient Hindu wisdom as a panacea to growing global materialism and societal fragmentation.

The Oliphant boys would later recall visits to their home by a parade of eclectic souls. Faithful to Theosophy's central belief that followers could be reincarnated in human form, these men and women gathered around Mark's hand-made kitchen table in spectral light, attempting to contact the departed. Their means of communication was a wooden planchette, fitted with a writing implement that would scratch and dance across the table top under the lightest touch from the wide-eyed participants, supposedly scribbling notes from the spirit world.

It was one of several episodes during Mark's developmental years that had helped galvanise his abhorrence of fundamentalist religious or political ideologies. It had put paid to any thoughts of following his father's wish for him to pursue a calling to the High Anglican priesthood – a wish that meant Mark had taken additional Latin classes at primary school.

During their flirtation with Theosophy, Beatrice and Baron Oliphant had also embraced vegetarian eating. When Mark and his younger brother Nigel visited the farm of a schoolfriend and witnessed pigs being slaughtered, the chilling squeals and gushing blood had traumatised them both to the extent that they too swore off meat for the remainder of their lives. 'Ever since I could make decisions for myself, I've been a vegetarian … because I do not want to kill things in order to remain alive,' Mark later explained. 'I think it's totally unnecessary.'[7]

Theosophy and vegetarianism might have set them apart, but in other ways, the Oliphants were no different from most families of the pre-war era. The bookshelves that Mark had expertly fashioned through his boyhood passion for woodworking held many of the same classics that had enthralled the Rutherford family. And while Ernest Rutherford's entrée to science was Balfour Stewart's *Physics* primer, Mark Oliphant had found inspiration in a more seminal text. Despite Baron's long-held hope that his first-born should follow a divine path, in the course of wider reading recommended by his mother, the boy happened upon Charles Darwin's *On the Origin of Species*.

'The final chapters described in simple language the fascinating story of Darwin's theory of evolution,' Mark recounted decades later. 'The story of Genesis suddenly became, for me, an allegory. This first introduction to the scientific method, its logic and its appeal to reason rather than to blind faith, was the turning point in my life.'[8]

Although Mark Oliphant's creativity had initially found expression in the perfectly proportioned tables, bookshelves and garden furniture he crafted from saplings sourced in the Adelaide Hills, his intuitive feel for practical science had likely been nurtured in the self-styled engineers' shed that Baron's younger brother assembled behind the senior Oliphants' family home in Dulwich. As a young boy, Mark had been captivated by the lengths of pipe, reels of wire, boxes of bits and bolts, and the regular parade of hard-bitten characters who would arrive nursing lifeless items, only to see them brought back to health after minor tinkering. Or through ferocious hammering.

Mark's own experiments began with dismantling clocks that he would then rebuild. By age twelve he had designed an electric alarm clock that he presented to his parents, along with responsibility for waking him in the mornings. He then moved on to more elaborate devices, often involving an electric charge administered by a hand-cranked magneto, or his potent home-made recipe for gunpowder.

At a school fête, he set up a sideshow scam that involved dropping a dozen pennies into a tub of water, through which he would discreetly send a low-voltage electric current. Unsuspecting punters were told they could keep as many of the coins as they could fish out once they had thrown their own penny in, from where they stood on a strategically placed damp burlap sack that acted as conductor. According to Keith Oliphant, another of Mark's younger brothers, it was a foolproof trick until one student, clad in rubber-soled boots that negated the shock, reached in and cleaned out their kitty.[9]

New Zealand, 1880s

Ernest Rutherford's first recorded science experiment was an amateur attempt to make a miniature battle cannon from the brass tube of a hat peg, outside the family home at Foxhill in the early 1880s. After filing a small hole in the tube's base to take a touch paper, he clamped the cylinder to a loosely constructed wooden frame and placed it twenty metres from its intended target. He poured a healthy dose of gunpowder into the tube, then jammed the marble that was to act as a cannonball inside too. So tightly was the marble packed that when the powder was lit, the brass casing blew apart. Mercifully, the only damage inflicted was to the splintered frame.[10]

Whereas Mark Oliphant gleaned practical skills from his uncle and his friends, Ernest Rutherford benefited from the insights of his artisan father. James Rutherford might not have received formal schooling, but he was conscientious and resourceful, and could produce, by his own hand, much of the equipment required for his milling operations and the family farming property. When first establishing his various enterprises around Foxhill, James was known to have travelled on a bicycle whose frame was constructed entirely from wood, a product of his own design and manufacture.

His son's interests were not quite so entrepreneurial, but often as methodical. Ernest Rutherford created working models of water wheels studiously copied from those that drove his father's mills. He too enjoyed dismantling clocks and studying their inner workings, and while his timber-turning skills were never as refined as Mark Oliphant's, the wooden potato masher Ernest lovingly carved as a kitchen aid for his grandmother was considered of such rare value that, after his death, it was displayed by London's Royal Society. Perhaps most remarkably for a boy of barely twelve, he built his own camera from parts he accumulated, and delighted in photographing daily life in rural New Zealand.

New Zealand, 1880s; South Australia, 1910s

Among the academic and pragmatic symmetries of Ernest Rutherford and Mark Oliphant's upbringings on either side of the millennium and the Tasman Sea, perhaps the strongest similarity was their respective relationships with nature. In large part, that affinity stemmed from the financial constraints within which both their families lived.

In lieu of exotic vacations and expensive toys, Rutherford and Oliphant – like so many children who fortuitously grew up removed from the soot and stench of Victorian-era industrial cities – found adventure, enlightenment and independence in the pristine bushland, limpid streams and abundant wildlife found just beyond their front doors. For both boys, the outdoors was a world of wonder that demanded close and repeated exploration.

The Oliphant boys were known to trek far afield, on expeditions led by Baron when he arrived from his city office job for weekend and holiday stints. To reconnect with life and family, and grant Beatrice some overdue respite, he would take his four eldest sons on epic walking adventures the length and breadth of the Adelaide Plains.

One of these followed the course of the Onkaparinga River that runs through Mylor. It was a journey of some 250 kilometres through mostly unmapped bushland, and it provided Mark with the most vivid memories of his treasured rustic upbringing.

> We carried swags and billies, and slept in the open on fragrant beds of gum leaves and bracken. The Mount Compass area was then just being developed for vegetable gardens. Near Clarendon, one could buy magnificent strawberries, and on the slopes below the road outside Willunga we picked the largest and tenderest mushrooms I have ever tasted. Mount Compass potatoes boiled in the billy or roasted in glowing embers to have crisp shells, split and plastered liberally with the very salty local butter, and eaten at dawn in the open, made a breakfast with royal qualities for hungry walkers.
> We bathed without hindrance of costumes in pools along the Onkaparinga and other streams, in the mouth of the River Murray and in the sea. The beauty of the Inman Valley, and the country between Cape Jervis and Yankalilla, with its heat-haze of eucalyptus on a summer's day, and glimpses of the cool sea, the road lined with dusty Christmas bush in flower, are memories of a wonderful boyhood.[11]

Mark Oliphant's connection to the quartzite-studded, baked-earth tracks that wound through the thickly wooded Adelaide Hills would remain inherent to his soul, regardless of where work and life subsequently took him.

In Rutherford's case, outdoor adventures during his New Zealand boyhood were sometimes undertaken on horseback, and often involved venturing to burbling creeks with his brothers, where they would spear eels and hook trout. He also became something of a small-gauge rifle marksman, and to supplement his family's food supplies would pick off wild pigeons attracted by the berries of the South Island's native miro trees.

As intrepid as their oneness with nature emboldened them to be, these two products of isolated rural communities also carried a keen awareness of the dangers inherent in country colonial living. But while Mark Oliphant's scrapes were usually salved with bandages and a stern talking-to, Ernest Rutherford's childhood was beset by the harrowing mishaps and desperate tragedies that were ever-present among frontier families.

While bathing in the Wai-iti River with his brother Jim during his primary school days at Foxhill, Ernest found himself in difficulty when swept out past his depth. He was rescued by his attentive younger sibling, who hauled Ernest to safety just as he began to disappear beneath the surface.

The next two boys in the Rutherford lineage were not so blessed. Percy, then aged nine, and Charles, eight, went sailing with an older friend on Pelorus Sound one afternoon in 1884. The young brothers both slipped overboard when the boat tipped in a wind shear, and although James Rutherford and his elder sons frantically scoured the shoreline for months after they disappeared, their bodies were never found.

The tragedy deeply affected Ernest, who had planned to be part of that sailing expedition until required to make an urgent delivery to his father's flax mill. Instead, he received the news of the double drowning on his return home, and then had to break it to his inconsolable mother.

Martha Rutherford's demeanour changed forever that afternoon. No longer able to find lightness in life, never again did she play the Broadwood piano she had previously polished so lovingly every day.

2

THE WORLD AWAITS

New Zealand and South Australia, 1885 to 1919

Ernest Rutherford's response to his brothers' deaths was less outwardly conspicuous than his mother's. In what would become a template applied to tragedies that awaited later in his life, he immersed himself in scholarly pursuits as refuge from the grief and sadness that enshrouded the family home.

The reality of life in small, unsophisticated rural communities throughout New Zealand at that time was that only half the eligible students stayed on at school beyond age thirteen. The rate of those who remained even longer, to gain secondary education, was around one child in fifty. Secondary schools were invariably privately run, and the fees were a luxury clearly beyond Ernest's already overworked father. If Ernest were to escape the fraught and fragile existence his forebears had endured, a scholarship stood as his only hope.

He engaged in after-hours tutoring from his teacher Jacob Reynolds in preparation for the Marlborough Scholarship Examinations, which he was to sit over two days in December 1885. But when the Marlborough results were announced, Ernest had finished runner-up, fifteen marks adrift of the prize's winner.

It meant that, as he neared his fifteenth birthday, he had no choice but to remain at Havelock school to take a second shot at earning a scholarship at year's end. Failing that, he seemed destined to take up a career in government bureaucracy, having also successfully completed the junior civil service entrance exam late in 1886. Or he could join his father in milling timber, and perhaps even muster a few head of cattle.

So driven was Ernest Rutherford to reach secondary school that he would rise at 5am in the bitter chill of the South Island winter to attend the extracurricular classes in Latin and algebra that the dedicated Reynolds offered. When the examination results were announced in December 1886, Rutherford had scored 580 out of a possible 600: top of the Marlborough Scholarship list.

His path to further learning was assured, at least for a few more years.

South Australia, 1914 to 1917

In his early teens, Mark Oliphant moved away from the Adelaide Hills with his family, and settled back into life in the outer suburbs. The aptitude for learning that Mark had shown at Mylor spurred Baron Oliphant to find rented accommodation at Mitcham in Adelaide's southern foothills, near a reputable government school at Unley. It was early 1914, and the political storm-front building in Europe was yet to deliver its direct impact on Australia.

Keith Oliphant, almost four years Mark's junior, would recall that a precedent was thus set: from then on, Oliphant family life would be moulded to enable Mark's academic ambitions.

> Mark seemed to sense things rather than learn them. He seemed to possess an instinctive mechanical insight.
>
> Our parents had to make real sacrifices for the sake of their sons' education, but Mark was so outstanding that that the rest of us had to wait our turn. Mark always gave me an inferiority complex. He was extremely unusual in his ability to understand things so quickly.
>
> Learning has never been a problem for Mark. The teachers would all say to me 'you're not like your brother'. Mind you, he was a dominant, arrogant boy, always full of self-confidence. You can't embarrass Mark by pointing out when he has made mistakes. He'll admit any mistakes readily, but it doesn't worry him.[1]

At Mitcham the family rented a large stone cottage, featuring a couple of basement nooks that made ideal retreats for a teenage boy with a penchant for inventing. While Mark quickly claimed those spaces for designing and building what were dubbed 'raggedy, baggedy engines'[2], a greater challenge was to deter his inquisitive brothers.

Although they usually padded about the house barefooted, Mark could sense when his siblings lurked outside the room's heavy wooden door. That was his cue to start hand-cranking his magneto, which he had sourced from a wall-mounted telephone cabinet rendered obsolete by Australia's recently federated telecommunications system. The small electrical charge that the magneto would deliver to the phone was instead directed to the wired-up door handle of Mark's room. Even with the obligatory damp sack laid outside as an unwelcome mat, the low conductivity of the brass doorknob meant sufficient current was administered to elicit a yelp and a retreat from a would-be intruder, but no risk of more serious injury.

Mark then set his sights on more ambitious experiments. Conspiring with a neighbouring youth, he turned to telegraphy. The friend's older brother was an electrician, which allowed Mark and his mate to borrow his tools to string up a single-wire telephone connection between their two houses, around 150 metres apart.

The most daring of Mark's devices were saved for Guy Fawkes Night each November. Having co-opted Keith to sneak into a nearby mansion one year and stealthily remove a few brass doorknobs, Mark transformed them into mini-grenades by filling them with gunpowder. He once boasted that his recipe for the chemical explosive was so volatile it was likely to explode 'if a fly walked over it'[3] – as he once proved, when he brushed his thumb into a pile being otherwise carefully prepared in his basement 'laboratory' and it ignited in his face. A similar misfire from a home-made 'throw down' bomb saw him attend school for a week with his head wrapped in bandages.

The mishaps did not dissuade the Oliphant boys from roaming the foothills of Mitcham, drilling holes in farmers' fence posts, filling the recesses with powder and detonating them via a length of touch paper. They would watch from a safe distance as the blast brought a cloud of splinters and the acrid burn of saltpetre.

New Zealand, 1887 to 1889

Courtesy of his Marlborough Scholarship, Ernest Rutherford arrived at Nelson College from Havelock at the start of 1887 to begin his two years of senior schooling. In itself, that represented a significant achievement for the son of a flax farmer in the still-developing colony.

Of the college's eighty enrolled pupils, aged from ten to twenty-one, two dozen were boarders. Each was assigned a place in one of the four-bed, draughty dormitories, with individual washing bowls the sole concession to creature comforts.

Not only had sixteen-year-old Ernest been thrust into a foreign existence, but he soon became even further removed from home. The need to find fresh sources of flax sent James Rutherford and his family to the Pungarehu area on New Zealand's North Island. The 500-kilometre journey from Nelson would mean a lengthy steamer voyage, so it would only be during end-of-semester breaks that Ernest could contemplate returning home.

He therefore threw himself, with country-boy vigour, into college life as a salve for homesickness. He became a hefty and wholehearted, if not especially silken, forward in the school rugby team, which held powerful significance, given that Nelson College was involved in the sport's first competitive match in New Zealand. However, it was his achievements in the classroom that convinced Rutherford his future lay in further study.

Having regularly topped grades and won prizes in mathematics, history, English literature, French and Latin during his first two years at Nelson College, he began to eye the prospect of undertaking a degree at the University of New Zealand in Christchurch, around 400 kilometres further south. Achieving that next academic goal would require exemplary grades in the matriculation exam to secure yet another scholarship. So the already driven student stepped up his study workload. He took additional evening classes, and his maths and science master noted in the end-of-1888 report that his pupil 'overhauled the work in shorter time than any boy I ever had'.[4]

After sitting his matriculation exam, Ernest began summer holidays with his family on the North Island. Weeks passed before he received confirmation that he was among the thirty-one students to have passed the exam with credit, and had therefore qualified for university admission.

However, his name was not among the ten who had secured bursaries to cover the cost of their courses. Given his father couldn't even afford a permanent dwelling at his Pungarehu mill site, with the rest of the family renting forty kilometres away at New Plymouth, he certainly wasn't able to subsidise his son's pursuit of tertiary education.

Once again, a heavy choice loomed for Ernest Rutherford. Although he had already sat and passed the New Zealand civil service entry test, he had foregone a bureaucratic cadetship in the hope of reaching university. Economic pragmatism now suggested that safe employment in an office would be his destiny after all.

But his heart still told him that his calling lay in science, and one faint light flickered in the middle distance.

Despite failing to secure a university scholarship, Rutherford was entitled to remain at Nelson College for an additional twelve months as the terms of his original three-year Marlborough bursary allowed. It meant he would essentially repeat the previous year's work, but could then compete for one of the ten Junior Scholarships on offer across the entirety of New Zealand. If he was successful in that even more competitive field, then a university place awaited.

When he returned to Nelson College as Head Boy in 1889, the extroverted manner that would later shape his leadership of the Cavendish first began to take form. His boisterous humour, booming laugh and unflagging commitment to every pursuit he tackled – whether after-school lessons, rugby or bouts of boxing – became the building blocks of his adult persona.

If there was one characteristic that stood out, it was the ability he had possessed since secondary school to concentrate single-mindedly regardless of tempest raging around him. At times, that chaos came from less diligent classmates who landed good-natured blows upon him as he studied in the library.

Rutherford's capacity to perform under all manner of pressures instilled within him crucial confidence when he again took the matriculation exam at the end of 1889: the final chance for the scholarship he needed. But past experience also prepared him for failure.

Before the exam results were known, he lodged an application for a vacant teaching post at New Plymouth High School, close to where his mother and siblings were living. To history's enduring benefit, he was rejected for that job, and he learned soon afterwards that he had placed fourth among those contesting the ten nationwide Junior Scholarships.

His path to the University of New Zealand at Christchurch, better known as Canterbury College, was clear.

South Australia, 1914 to 1918

Six months into Mark Oliphant's first year at Unley High School, the school's diminutive headmaster, Major Benjamin Gates, appeared at the classroom door to announce that Australia, by dint of its imperial ties to Great Britain, was at war with Germany. While the gravity of this pronouncement was immediately obvious to his parents, Mark would remember the onset of war and the subsequent campaigns as something of a distant adventure.

The most obvious impact on his life was that, over the next year or more, many of the school's male staff enlisted to the cause. They were mostly replaced by female teachers, some of whom had been hastily briefed on the curricula they were to deliver. Mark Oliphant's second-year French teacher blushed as she confessed her limited knowledge of the language, her candour earning forgiveness from the class as she stumbled over Gallic pronunciations.

As Mark's secondary education progressed against the backdrop of the Great War, the arduous Latin classes he had been compelled to complete at primary school in readiness for theological studies prepared him well for senior school exams. By then, however, Mark's belief that science was his future had only strengthened.

'The wonders of chemistry made me wonder whether I could satisfy my family's "do-gooder" feelings by becoming a doctor, who used chemistry in treatments,' he later recalled, citing his parents' attachment to the Anglican faith that had remained devout, even during their dalliance with Theosophy.[5]

To realise that aspiration he would need to sit the fifth-year Senior Public Examination which, in 1917, Unley High did not offer. So, with his mother's firm hand at his back, he transferred to South Australia's foremost government secondary institution, Adelaide High School, located almost ten kilometres away in the city's centre.

New Zealand, 1890 to 1892

Despite its frontier location, Canterbury College – like the settlement of Christchurch itself, planned under the systemised colonisation model – aspired to exude the very best qualities of British life. It was founded as an Antipodean bastion of Oxbridge education, and as Rutherford immediately noticed upon arrival, it strictly enforced the wearing of academic gowns and mortar-board headpieces by campus students. Frivolities such as smoking and whistling were summarily forbidden.

At the time of Ernest Rutherford's arrival in 1890, the institution catered to around 150 full-time students and employed just five professors. Among the latter cohort was Alexander Bickerton, Canterbury's inaugural chair of chemistry.

Professor Bickerton was as much a showman as a scholar. His theatrical classroom demonstrations routinely featured flashes of light and thunderous explosions that entranced Rutherford who, until that time, had felt more comfortable studying theory.

That preference was made clear early in his undergraduate career. In his first year at Canterbury College, Rutherford was adjudged co-winner of the annual mathematics exhibition prize. Tellingly, in addition to the initial study load for his three-year bachelor degree, which included French, Latin and applied mathematics, he signed up for a weekly Saturday-morning 'Physics for School Teachers' course, keeping open the option of a job in the education system.

But as he worked industriously through his degree studies, it was the thrill of experimental work uncorked by Professor Bickerton that began to dominate his time and thinking.

South Australia, 1918

Soon after the guns finally fell silent and Europe confronted the challenges of post-war peace, Mark Oliphant completed his secondary schooling and faced a similarly uncertain future. In later life, he would concede he was no natural scholar. He blamed this partly on the deficiencies in his sight and hearing, and partly on teachers whom he pejoratively labelled 'mumblers'. He also recognised that, as a teenager, he felt more fulfilled when he was 'fooling about in the shed' than when leafing through textbooks.

'I was never a good student,' he would later admit, with a shrug and a smile. 'I was never top of class or anything of that sort. I was always in the middle range.'[6] By his own admission, he struggled with mathematics, a drawback for any aspiring physicist.

One tepid school report from midway through his first – and only – year at Adelaide High School in 1918 noted 'he is keenly interested in physics and chemistry and doing good work', alongside a physics assessment of fifty-six per cent. 'He has maintained his interest in his work and is very efficient on the practical side,' assessed another later in the year, despite Oliphant's being injured in a laboratory mishap involving hydrochloric acid that left him sporting more facial bandages and an arm sling.

Where Oliphant revealed himself to be undeniably gifted was in his capacity to design and construct apparatus to bolster the equipment supply at his resource-strapped high schools. His interest in laboratory equipment deepened during his final year at Adelaide High School, where Victorian-era instruments of lacquered brass and polished mahogany glinted from within a glass-fronted display cabinet. Oliphant became obsessed with an old twin-cylinder vacuum pump, which he would religiously clean and grease to improve its function – and was duly castigated by his physics teacher and headmaster for indulging in such an abject waste of time.

Other pieces of apparatus took shape in his experimental den beneath the family home. These were not mere 'raggedy, baggedy engines', but often intricate devices, including voltameters, gas absorption tubes, mercury air pumps, induction coils, electrometers, galvanometers, hydraulic aspirators, thermometers, glass plates, organ pipes, sirens and an automatic tuning fork.[7]

A science career clearly beckoned, but it would remain a dream if Oliphant's examination results were not exceptional enough to secure one of the twelve scholarships on offer to aspiring secondary students throughout South Australia at war's end. When results were posted at the conclusion of the 1918 school year, Mark's showed he had matriculated in English literature, arithmetic, algebra, Latin, trigonometry, physics, chemistry (with credit), physical geography and geology. A scholarship, however, had eluded him.

So Mark Oliphant left Adelaide High aged seventeen, with no option other than to find a job. To aid that endeavour, he held a matriculation certificate and a letter of support from the school's headmaster, who recommended his pupil as 'a lad with undoubted bias towards science, of gentlemanly address, good mental powers and thoroughly straightforward and trustworthy'.[8]

What Oliphant lacked, however, was any demonstrable work history. This meant that unskilled office jobs became his starting point, and the civil servant existence that had entrapped his father loomed forebodingly.

New Zealand, 1891 to 1893

During his first year at Canterbury College, Rutherford's principal academic rival was his boarding house roommate Willie Marris, who would subsequently achieve fame as governor of the eastern Indian province of Assam. Marris maintained that Rutherford was a superior mathematician at university, but that he (Marris) achieved marginally better end-of-year marks because Rutherford suffered anxiety under the pressure of exams.

That was doubtless a legacy of his chequered history chasing scholarships, but by the completion of his second year at university – when students were required to sit their first round of Bachelor of Arts exams – Rutherford's confidence in his abilities had grown demonstrably. Come the end of 1891, he had completed the first part of his BA degree by successfully sitting exams in physics, pure mathematics, applied mathematics, English and Latin. A year later, he completed a clean sweep of those subjects as well as French to secure his undergraduate degree.

In his final undergraduate year, Ernest Rutherford also took part in New Zealand's senior scholarship exams for both mathematics and physics. When the final marks were assessed, the physics scholarship was awarded to a rival student from Otago by such a narrow margin that the examiner, based in England, recommended that a new precedent be created by awarding *two* prizes. The scholarship board advised that that would be unnecessary, as Rutherford had finished top of the pool in mathematics, and was therefore assured of receiving a bursary.

Thus guaranteed short-term funding to continue his academic career for a further year (of honours or master's study), the twenty-one-year-old became one of only fourteen postgraduate research students in New Zealand in 1893 – seven of whom were at Canterbury College. While the other six pursued higher degrees in languages, Rutherford's ambition was obvious from the six courses in which he enrolled – honours maths, honours chemistry, chemical laboratory practice, practical physics, general biology and senior botany.

So unusual was his choice of postgraduate subjects that he was the lone member of his mathematics class. The results that he achieved also set him apart, and by year's end he had attained a Master of Arts qualification with a rare double distinction: first-class honours in mathematics and physical sciences.

Despite his glittering academic record, Ernest Rutherford found employment opportunities limited as New Zealand's economy continued to feel the impact of depressed prices for its major export commodity, wool. So, with the guarantee of modest income from tutorial work if he remained at university, Rutherford enrolled in a Bachelor of Science program that included courses in geology and chemistry.

Even though full-time work eluded him, the once-reticent, solitary schoolboy had gained such assurance from his academic achievements that he not only signed up for the college dialectic (debating) society, but also volunteered to serve as one of its office-holders. His seniority was recognised within Canterbury College's rugby club too, where he held the office of assistant secretary. While his football acumen fell well short of his intellectual prowess, Rutherford was part of the college's first team to play before a paying crowd, beginning the 1893 season at Christchurch's premier sporting venue, Lancaster Park. His reputation as a scholar, and even as a sportsman, continued to grow.

While his mathematics scholarship had enabled him to continue studying during his master's year, it did not cover the cost of board at Canterbury College. Compelled to find private lodgings in Christchurch, sometime before 1893 Rutherford took a room in the home of widow Mary Newton at Carlton Mill Corner, a ten-minute walk from the campus. Mary Newton's husband had effectively drunk himself to death five years earlier, leaving her with three young boys and a teenage girl, also christened Mary but known to most as May.

As daughter of Christchurch's inaugural town clerk, Mary Newton had been afforded a good education and a level of social standing, and both of these privileges were mirrored in her eldest child. The aloofness young May would sometimes display towards others reflected her mother's former societal status, and the teenager's strong moral code had been honed by her mother's staunch anti-smoking and anti-drinking philosophy, a direct response to her late husband's alcoholism.

However, as the eldest daughter of a young (forty-one-year-old) widow, May was also regularly called upon to care for her three little brothers, William, George and Charles. She then won prizes during her senior schooling – including awards in mathematics and science that prompted some to suspect she was receiving private tuition from the family's boarder, almost five years her senior. Aged sixteen and having lost her father before she reached adolescence, May had forged a close bond with the garrulous if studious young man of the house.

Just how quickly May's romantic relationship with Ernest Rutherford flourished under Mary's hawkish vigilance is not known, but when the boarder returned to his family at Pungarehu for the summer holidays at the end of 1894, he was accompanied by his landlady's then eighteen-year-old daughter.

May's arrival on the North Island created ripples within the Rutherford household. Ernest's five sisters found her to be snooty and spoiled. They resented the fact that May was able to accompany Ernest on his regular horse-and-trap journeys to collect mail from the nearby town's post office, yet they were never invited. Ernest's girlfriend was decidedly more popular with his brothers, and she once coquettishly told James Rutherford: 'If I cannot have your Ern then I'll have one of your other sons.'[9]

However, according to the custom of the time, May's presence at Pungarehu implicitly announced that the couple was engaged to be wed, as soon as Ernest had secured suitable means. Towards the end of 1894, Rutherford had served an inauspicious stint as relief teacher at Christchurch Boys' High School as he explored the possibility of a teaching career once his Bachelor of Science was completed. But his impatience with pupils who did not share his logical insights and grasp of difficult scientific subject matter meant that neither teacher nor students had gained much from the experience.

'He was entirely hopeless as a school master,' one of those disillusioned pupils would write of Rutherford the teacher. 'Disorder prevailed in his classes … I do not remember myself following any of his intellectual processes on the blackboard. They were done like lightning.'[10]

At the start of 1895, Rutherford therefore returned to Canterbury College, with May. Their future plans were beset by uncertainty. May had enrolled to study English, French, Latin and zoology at Canterbury College; she would finish the year having failed all four. Whether her academic results reflected the turmoil of the couple's circumstances or confirmed that her earlier successes were built largely on the expert input of her family's household boarder is not known.

Her 'fiancé', meanwhile, took up tutoring to help defray the cost of his ongoing studies. He also began his first forays into dedicated research work, although he never explained his rationale for pursuing such matters as magnetic viscosity and properties of electromagnetism (as per his first two published papers in 1894 and 1895). It was a field of endeavour that also led Rutherford to explore new means of producing and detecting radio waves, seemingly in line with the great scientific quest of 1895 (also pursued at that time by Italy's Guglielmo Marconi).

What was abundantly clear, however, was that farming on the North Island would be Rutherford's default career if another scholarship could not be won to keep him in science.

South Australia, 1918 to 1919

It was connections rather than qualifications that landed Mark Oliphant paid work when he completed secondary school in late 1918. Schlanck and Co., the Adelaide manufacturing jeweller that hired the seventeen-year-old as a general hand, was run by a friend of Baron's. Established four decades earlier by Salis Schlanck, a Jewish émigré from Prussia, the firm had won government contracts in 1918 to produce silver-plated peace medals as mementos of gratitude for those who had served in the Great War.

Mark's place in this enterprise was to labour over the blacksmith's forge where the metallic alloys were melted down, and sweep the soot-coated floors of the sweltering workshop. While it would instil in him a lifelong interest in silversmithing – the hefty sterling candlesticks he made and bequeathed to South Australia's Government House being perhaps his most ostentatious pieces – his distaste for this unskilled work stoked a need to chase more rewarding roles.

It was a pursuit that proved repeatedly fruitless.

He first applied for a job as assistant in the testing department of the Adelaide Cement Company, where the interviewer bluntly told the boy that his ambitions were too lofty. Oliphant preferred to believe he had been rebuffed due to his clothing, mostly home-sewn by his mother, who produced her family's entire wardrobes save for their boots.

He then tried for a similarly junior position at the fledgling Commonwealth Institute of Science and Industry. This time he missed out because the institute was deemed not yet to constitute a permanent statutory authority, and was therefore unable to hire and train staff.

The institution that did see a place for the tall, assured eighteen-year-old was the South Australian Public Library, which had expanded beyond the Institute Building that sat alongside the now-closed destitute asylum. Largely oblivious to the colonial significance of books with

'Circulating Library' embossed upon their front covers, Oliphant quickly grew impatient with the menial chores of hand-writing catalogue cards in elaborate copperplate, brand-stamping newly arrived books and re-shelving older ones. More than once, he was upbraided for wasting work time reading texts he was supposed to be restocking.

However, the mundane clerical work came with a crucial benefit. Thanks to encouragement provided to library 'cadets', Oliphant was able to pursue a limited range of study options at the neighbouring University of Adelaide as an adjunct to his employment.

It was through this scheme that Mark Oliphant enrolled in a general science degree involving some chemistry, some physics and a smattering of mathematics, which he studied through an 'evening studentship'. The aspiring science scholar had found a route to further learning.

New Zealand, 1894 to 1895

During his fourth and then fifth year at Canterbury College – studying for a Master of Arts and then a Bachelor of Science degree – Ernest Rutherford's scientific ambitions took a decisive turn. He had successfully evolved from a diligent mathematics student into an accomplished researcher, but in order to gain a master's degree he was required to undertake an original investigation.

Professor Bickerton suggested that he study the prospects of forming organic molecules by passing electrical discharges through inorganic gases such as carbon dioxide. But Rutherford momentously chose to turn his back on chemistry, and instead pursue physics.

Initially that shift came through a continuation of work he had begun as an undergraduate, on a topic that excited huge interest in science around the turn of the twentieth century: the properties of electromagnetic (specifically radio) waves. He secured the use of a basement room beneath the steeply raked floor of a lecture theatre at Canterbury College that was otherwise employed to store ceremonial gowns, and transformed it into his own experimental den in lieu of a dedicated physics laboratory on campus. And it was in this unlikely, unused cellar that Rutherford's brilliance exploded into full flower.

The experiments he conducted with radio waves, using a Hertz oscillator he constructed on the concrete floor of the frigid store room, entertained fellow students, who gathered to watch him send and then record signals through the building's thick stone walls and iron door. His peers also marvelled at the technical skill he had clearly learned in his father's home workshops, which was now being applied to ever more intricate machinery.

However, it was his earlier 1894 paper exploring the magnetisation properties of iron when subjected to electric current oscillated at high frequencies – sometimes flicking on and off at intervals of one-hundred thousandth of a second – that gained him far wider attention. This was due not only to the sophistication of the equipment he devised in such comparatively primitive surrounds, which he then applied through his elegantly simple methodology, but also to the audacity of the findings in his final published reports.

The paper that Ernest Rutherford presented to the Philosophical Institute of Canterbury on 7 November 1894 challenged the previously published conclusions of no lesser a triumvirate than Oliver Lodge, then Professor of Physics and Mathematics at Liverpool's University College; J.J. (Joseph John) Thomson, Cambridge University's Cavendish Professor of Physics; and the recently deceased Professor Heinrich Hertz of the Bonn Physics Institute, whose name would become immortalised as the unit of measurement for radio waves.

In his first published works, Rutherford effectively declared the exalted trio's shared view that iron was not magnetic at very high frequencies to be fundamentally wrong. While the twenty-three-year-old student's assertions were clinical rather than boastful, the assuredness shown by an upstart research student from the far side of the globe caught the attention of more than a few in the closed shop of Europe's science establishment.

The secret behind Rutherford's success in proving results that contradicted the learned physics minds of Britain and the Continent was his ability to produce apparatus capable of providing minuscule measurements. This was a feat beyond the scope of most other scientists at that time. This newly realised practical aptitude was complemented by the comparative technical simplicity of the experiments he devised, and the absolute immersion he brought to his chosen subject. Over the course of a couple of published papers, a potent new force in the field of physics was announced.

It was clear, however, that prominence in respected scientific journals and satisfaction at shaking up the institutional establishment as a maverick researcher from the new world were not Rutherford's principal motivations. Apart from the sheer joy he gained from plotting experiments and obtaining results, his main ambition was to win yet another fiercely contested scholarship. The prize he sought was one born of Britain's bold gambit to re-establish its centres of learning as the yardstick for worldwide scientific research, and one that Rutherford hoped would grant him access to the very institutions, including Cambridge University, that he had dared to ruffle.

The 1851 Research Fellowships were a legacy of the world's first international fair, London's Great Exhibition. It had unfurled between late spring and early autumn of that year at Hyde Park, within an overgrown greenhouse designed by Joseph Paxton that would later earn more regal recognition as the Crystal Palace.

Under the patronage of Queen Victoria and her consort, Prince Albert, the exhibition trumpeted Britain's industrial prowess and the scope of its empire, showcasing technological enterprise and cultural curiosities from almost fifty sovereign states. Such interest was aroused that around 6 million visitors – a third of Great Britain's total population at the time – passed through the shimmering glass-and-steel cathedral's doors.

That patronage ensured that the event turned a profit of £186,000: around £24 million in the currency of today. Much of that windfall went to establishing a precinct immediately south of Hyde Park that became, and remains, home to some of London's essential landmarks – the Victoria and Albert (displaying many of the exhibition's original items), Science and Natural History Museums.

In addition, Prince Albert oversaw the establishment of a scholarship fund that would allow the most promising research students from across the Empire to pursue their studies at Britain's acclaimed centres of learning. It was proposed that each year, eight '1851 Research Fellowships' would be offered to worthy aspirants throughout the United Kingdom, as well as Britain's colonies, which would be eligible to submit candidates according to a roster.

It took until 1891, three decades after Albert's death, for the first of these scholarships to be realised. That timeline, which saw New Zealand able to put forward candidates on a biennial basis, coincided neatly with Ernest Rutherford's broadening ambitions and would ultimately benefit Mark Oliphant as well.

But this wasn't the only item of historical coincidence to fall in Rutherford's favour. At the time of the Great Exhibition, Cambridge and Oxford Universities were accessible only to Britons. It took until 1895 – just months after Rutherford submitted his initial scholarship application – for Cambridge to accept postgraduate students who had completed their undergraduate degrees at other institutions, whether in Britain or beyond.

On the basis of his landmark thesis work at Canterbury College and the notice it had gained in the wider science community, Ernest Rutherford clearly fitted the 1851 scheme's criteria. His only obvious impediment was his historically poor record at winning scholarships at the first attempt.

His odds narrowed when only one rival emerged: a chemistry student from Auckland named James Maclaurin. The University of New Zealand – now disestablished but then the nation's over-arching and sole degree-granting body that incorporated colleges in Auckland and Christchurch – forwarded both applications to the 1851 Commission in London. That was where, after several months of detailed scrutiny and some direct lobbying on Rutherford's behalf, the decision was made. New Zealand's allocated scholarship for 1895 was Maclaurin's.

The ruling led some of Rutherford's more indignant academic supporters to claim precedent should be shelved and a second scholarship awarded, such was their conviction in his case. But that suggestion was rejected by the commission. With his chances of finding further patronage ever shrinking, Rutherford was all but convinced his days as a scientist were over.

History might have read very differently if the paperwork sent to confirm Maclaurin's scholarship had not raised immediate concern in

Auckland. The clear stipulation that the recipient remain in full-time study for the bequest's two-year duration, and not engage in any paid employment throughout that time, was at odds with Maclaurin's plans. Newly married and holding a lucrative job as a government analyst, he knew he would not be granted two years' leave from a position he could not risk vacating. In April 1895, Maclaurin declined the offer. So the University of New Zealand turned to its fall-back candidate.

While the machinations of bureaucracy ground slowly on, Rutherford had returned to the family property at Pungarehu to take a break from his research and tutoring work during the mid-term holidays that included the Easter weekend. It was there, as winter approached and he was turning sods in the family's potato patch, that a telegram was delivered to inform him the 1851 scholarship for that year was his.

The tale Rutherford would later recount to Mark Oliphant was that when his mother read the cable aloud, he threw down his shovel and roared: 'That's the last potato I'll dig.'[11]

New Zealand to England, 1895

Just a few months after receiving news of his scholarship in July 1895, Ernest Rutherford sailed from New Plymouth aboard the SS *Takapuna*, bound for Cambridge University's Cavendish Laboratory.

Rutherford's 1851 exhibition scholarship might have entitled him to tuition at any university that would accept him as a research student, but it did not extend to passage fare from New Zealand. Although he had three university degrees to his name – his research into the capacity of high-frequency electrical discharges to magnetise iron having earned him a BSc in the course of a single year – his attempts to find paid work had proven consistently unsuccessful. Therefore he had been compelled to borrow funds for his journey (his 'grub stake', as he called it) from his eldest brother, George, now a prosperous North Island farmer.

Rutherford's choice of Cambridge as his preferred research destination was primarily driven by the chance to work with the incumbent Cavendish Professor, renowned physicist J.J. Thomson. While Rutherford's early research paper might have challenged some conclusions Thomson had reached, the young research student held lofty regard for the reputation and rigour of the humble professor, who would forge an early path in sub-atomic study.

But before introducing himself to the most eminent physics researcher in Britain, Rutherford planned to meet with the man who held that title in the Antipodes: William Henry Bragg, at the University of Adelaide. Thus he fleetingly broke his voyage in South Australia during the last month of 1895's southern winter.

Bragg had garnered acclaim over the preceding decade as Elder Professor of Pure and Applied Mathematics, and shared Rutherford's interest in radio transmission. Moreover, he was himself an alumnus of the Cavendish Laboratory. It was in fact J.J. Thomson who had encouraged Bragg to declare his interest in the vacant professorship in distant Adelaide: a bold gambit at twenty-three, the same callow age at which Rutherford was now beginning his journey across the globe. Bragg had achieved much since departing his native England — not least his marriage to Gwendoline, daughter of Adelaide postmaster Charles Todd, the employer of Baron Oliphant's father Harold.

Rutherford knew no-one who lived outside his homeland. And those of whom he was aware through their reputations were mostly in an alternative hemisphere, if not an altogether different league. So it was as much to glean insights as to who and what awaited at Cambridge as to compare notes with Bragg that led Rutherford to Adelaide University in August 1895. There he would encounter, not for the last time, a kindred spirit who would prove a lifelong friend.

When Rutherford returned to the waiting steamer hours later, he carried a formal letter of introduction from Professor Thomson's former student, as well as some fresh ideas about 'radioactivity' — the term then applied to wireless telegraphy.

Throughout his next fifty days at sea, Ernest Rutherford frequently mulled over the material Bragg had given him. It helped to quell the nervous excitement coursing through him while the distant dream he had chased since schooldays drew closer, then closer still, with each hour that passed upon the vast, heaving Indian Ocean.

3

'RABBIT FROM THE ANTIPODES'

Cambridge, 1895 to 1898

Ernest Rutherford's mood was as bleak as the low, grey sky on that dull, chilly Thursday morning. It was early October 1895, and he was on his way to London's Liverpool Street Station, having hauled himself from his boarding-room bed, where he had spent the previous three days with a severe cold compounded by a painful bout of neuralgia. Now he was forced to push through the crush of commuters huddled against the early autumn cold and the Victorian-era grime, while sporting a pronounced limp.

As if his first days in the unfamiliar and unwelcoming city had not sufficiently stirred pangs of homesickness for the unfettered streets and clean air of Christchurch, he had also slipped on a banana skin discarded on the footpath and severely wrenched his knee. If not for the gravity of his mission – a first meeting with Professor J.J. Thomson at the Cavendish Laboratory – the young visitor would have gladly returned to the refuge of his room.

However, the hour-long, non-stop journey flashed by in a montage of inner-city tenements, outer suburbs and – finally, as Rutherford's renowned good humour gradually returned – the deep green pastures of the eastern Home Counties. 'The country is pretty enough,' he later noted, 'but rather monotonous.'[1]

The first curiosity Ernest Rutherford observed upon alighting at Cambridge was that the railway station, with its grand archways and columns, decorated with ornate friezes and cornices, was not located anywhere near his ultimate destination. Fearing the impact that hordes of visitors might wreak on their staid community as rail networks spread across Britain in the mid-1800s, the influential colleges of Cambridge University had flexed their collective muscle and demanded the train station be sited two kilometres from the town's centre. And that no rail traffic be permitted to stop there on Sundays.

Hobbled and heavy-headed, Rutherford was therefore compelled to hail a horse-drawn cab to make his appointed meeting. The unexpected sights that confronted him once they cleared the low rows of railway cottages along Mill Road made him feel like he had travelled much further afield than the ninety kilometres that separate Cambridge from London.

As the cab skirted the large expanse of lawn that his driver proudly announced to be 'Parker's Piece', Rutherford glimpsed a few similarities with the world he had left behind at Canterbury College. One was the predominant dress code of full academic gown and mortar-board headwear for the young men, who gathered in groups upon the grass, or dashed down the winding streets and narrow alleys as the cab pushed deeper into town. What he had not expected was the preponderance of people on bicycles, a mode of transport that dominated the landscape in every direction his head swivelled.

Against the backdrop of the centuries-old limestone buildings and the haywire network of roughly cobbled laneways, Rutherford noticed other cultural peculiarities that he would come to understand more acutely over coming months. Among them was the edict that cap and gown be worn by students when out at night, on Sundays (except for the afternoon), during lectures, at chapel and in the dining hall, where college residents were expected to take meals at least five nights per week.

From the cohort of college fellows a number of proctors were elected to patrol Cambridge's streets at nights, and on Sundays, in full academic dress. As Rutherford would later understand and occasionally witness, these vigilantes were deployed to keep order by ensuring that the strict demarcation between senior and junior students – in addition to the stringent dress protocols – was observed. They were also authorised to issue monetary fines and administer discipline, a task they outsourced to accompanying pairs of constables, readily identifiable in top hats and tail coats. These 'bulldogs', as they were known, would dispense rough, ad hoc justice to students deemed to be misbehaving or inappropriately attired, or found to be consorting with young women.

Rutherford felt relieved when his driver announced they had reached the Cavendish Laboratory – only to discover he had been mistakenly deposited at Cavendish College, on the opposite bank of the River Cam.

Life in Cambridge, as he had gleaned in less than an hour in his new home town, loomed as both daunting and different.

* * *

As alien as Cambridge appeared and felt to Rutherford during that initial day trip, he took comfort from the ensuing meeting with J.J. Thomson, who would become his confidant and collaborator in many triumphs to follow.

Of that first meeting in the early autumn of 1895, Rutherford wrote to his betrothed Mary Newton, who assumed her given first name when there existed minimal risk of confusion with her namesake mother and was waiting patiently in Christchurch:

> I went to the [Cavendish] Lab and saw Thomson and had a good long talk with him. He is very pleasant in conversation and is not fossilised at all. We discussed matters in general and research work and he seemed pleased with what I was going to do.
>
> He asked me up to lunch … where I saw his wife, a tall, dark woman, rather sallow in complexion but very talkative and affable. Stayed an hour or so after dinner then went back to town [London] again. I like Mr and Mrs both very much. She tried to make me feel at home as much as possible …[2]

Thomson, who had been chosen as Cavendish Professor eleven years earlier at just twenty-eight, was the third man to hold that exalted office, a lineage that stretched back to the year Rutherford was born.

It was in mid-1871 that the Cambridge University Council had gratefully accepted a gift of £6300 (worth around £800,000 today) from William Cavendish, the seventh Duke of Devonshire, who had succeeded Prince Albert as the university's chancellor. The bequest was intended to bankroll the building and staffing of an experimental physics department. It was generally acknowledged that Cambridge's influence in that discipline had steadily waned since Sir Isaac Newton's discoveries in motion, optics and mathematics two centuries earlier.

The man appointed as the laboratory's inaugural professor was James Clerk Maxwell. His discoveries in relation to colour theory, thermodynamics and mathematics paved the way for enhanced understanding of the physical world, and the resultant upsurge in experimental and theoretical physics. He was also a pioneer of colour photography. Albert Einstein, when visiting Cambridge University during the 1920s, was congratulated on his remarkable discoveries but reminded that he effectively stood upon the shoulders of Isaac Newton. 'No', Einstein countered. 'I stand on the shoulders of James Clerk Maxwell.'[3]

Maxwell oversaw every detail of the proposed new Cavendish Laboratory, including its site in the rather grandly named Free School Lane, a narrow alleyway that skirts the rear wall of Corpus Christi College. This unobtrusive location was based on Maxwell's belief that the distance from Cambridge's heavily trafficked thoroughfares would minimise vibrations that might otherwise interfere with experimental work.

Incorporated into the three-storey building, within its starkly Victorian Gothic façade, was a ground-floor magnetism room, which was vital to the accurate calibration of all electrical equipment in the laboratory and set on a concrete slab almost fifty centimetres thick to ensure its stability. There was a 4.5-metre-high ceiling in the battery room, which provided constant electrical power to all demonstration and work spaces, and was located immediately below the lecture theatre. This theatre fitted 180 students, in steeply raked rows of seats, overlooking a massive oak demonstration bench. To overcome the chill that rarely abated during term time, warmth came from a network of hot-water pipes, which were fashioned from copper, so as not to play havoc with the magnets. And the sprawling first-floor laboratory, with its ten purpose-built work tables, allowed constant monitoring via two hatches secreted in the walls of the adjacent professor's room.

The laboratory was still three years from completion when Maxwell delivered his maiden address as Cavendish Professor on 25 October 1871. He articulated a clear vision of what he expected his new facilities to achieve:

> The familiar apparatus of pen, ink and paper will no longer be sufficient for us, and we shall require more room than that afforded by a seat at a desk, and a wider area than that of the blackboard. Whatever be the character in other respects of the experiments which we hope hereafter to conduct, the material facilities for their full development will be upon a scale which has not hitherto been surpassed.[4]

Over the next almost quarter of a century, the Cavendish Laboratory established and then embellished its reputation for diligent and important research. That growth was overseen by Maxwell until his premature death aged forty-eight in 1879, from abdominal cancer. The role of Cavendish Professor was then filled by John William Strutt, the third Lord Rayleigh, whose renown in experimental and theoretical physics was known from his earlier years at Cambridge. He had a personal interest in the determination of electrical standards, and would establish the ohm as the absolute standard unit of electrical resistance.

Under Rayleigh's guidance and his talent for raising funds from influential friends, the Cavendish widened its scientific ambitions to encompass the full breadth of classical physics. The laboratory became recognised as the benchmark against which all other experimental physics facilities could be measured.

It continued to grow in importance and repute under the stewardship of Thomson, whose research as Cavendish Professor focused on the mathematical and experimental issues of electromagnetism. That work built substantially on earlier discoveries by James Clerk Maxwell, and was a central factor in Ernest Rutherford's choice to undertake his 1851 exhibition scholarship at Cambridge.

* * *

J.J. Thomson immediately became more than a scientific mentor to his young New Zealand research student. He schooled Rutherford in Cambridge's mysterious ways, and inducted him into the curious rituals of Sunday golf. Thomson's wife, Rose, became almost a foster mother to the twenty-four-year-old, not only keeping a keen eye on his welfare, but also securing him lodgings with a kindly Cambridge widow. To stave off homesickness, Rutherford fixed photographs of familiar Christchurch landmarks to the walls of his rented room.

Within weeks of his arrival at Cambridge, his insatiable work ethic and unflinching focus had him spending five nights of every seven working in the laboratory, often until near midnight. Barely two months into his scholarship, Rutherford fronted his first lecture as a demonstrator.

Rutherford's dedication to science might have immediately endeared him to Thomson, but at an institution that had not previously hosted students from outside its own clearly defined stratum, he was generally greeted with suspicion. That grew to distaste among a coterie of postgraduate researchers who took a condescending view of colleagues who hailed from the further reaches of the globe, such as the United States, Canada, and certainly New Zealand. If the hale and hearty, egalitarian manner instilled by Rutherford's upbringing in a diaspora all but devoid of entrenched social class didn't immediately set him apart, then his booming voice, thunderous laugh and florid farm-boy appearance provided circumstantial cause for distrust. Invited to take a place at the head table during a formal King's College dinner in January 1896, Rutherford confided in a letter to Mary Newton that he felt 'like an ass in lion's skin'.[5]

One of his Cavendish contemporaries, Dr Andrew Balfour, wrote of him: 'We've got a rabbit here from the Antipodes, and he's burrowing mighty deep.'[6] Frenchman Paul Langevin, another of Cambridge's inaugural intake of overseas postgraduates, worked in an adjoining room to Rutherford's and was asked about their relationship. He replied bluntly: 'One can hardly speak of being friendly with a force of nature.'[7] One less diplomatic colleague reputedly referred to the colonial as 'a savage, however noble'.[8]

The disconnect was at times mutual as Rutherford struggled to make sense of Cambridge, and Britain more broadly. Even as summer brought out daffodils along the River Cam and picnic blankets on the meadows between the stately college buildings and the waterway – known collectively as 'The Backs' – he continued to find the soot and smoke that enveloped Victorian England towns utterly depressing compared with the crisp air of the Canterbury Plains. Travelling further into the Midlands, he was struck by the backwardness of British rural life, where thatch was still employed as a common roofing material. With disbelief, he wrote to his mother: 'You can't imagine how slow-moving, slow-thinking the English villager is. He is very different to anything one gets hold of in the colonies.'[9]

There was, however, familiarity to be found within the laboratory. Rutherford had resumed the research into radio waves that had captured his interest at Canterbury College. A few months into his scholarship, he presented a demonstration at the Cavendish during which the signal he sent from a Hertz oscillator (spark transmitter) was received almost 200 metres away. Months later, he set up the transmitter in an open space on the southern bank of the Cam and placed a receiving device inside a house on nearby Park Parade, where it successfully recorded the signal from a distance of 275 metres. Within days, he had extended that range to around 1.2 kilometres, double the distance across which Italian Guglielmo Marconi had recently been able to perfect transmission at his father's estate in Bologna. Consequently, Rutherford became the holder of the global benchmark in the lucrative developing technology of wireless transmission, if only fleetingly.

It was during 1896, and less than a year into his time at the Cavendish, that Rutherford abruptly quit radio waves as a research interest, thereby turning over the field of wireless technology – along with its accompanying fame and riches – to Marconi. While Rutherford's innate belief that science should be pursued for its pure outcomes rather than possible commercial gain has been suggested as a possible reason for the switch, it's equally plausible he had simply locked on to a fresh challenge. That challenge was the study of sub-atomic matter, which had become world physics' *cause célèbre* at almost the same moment as Rutherford entered the Cavendish.

***

In the final weeks of 1895, news had emerged from central Germany that drastically changed the landscape of physical science. Dr Wilhelm Röntgen, Director of Physics at Bavaria's University of Würzburg, had made a chance discovery when working with the sort of discharge tube that had become an instrument of choice for physicists. These were thin-walled, purpose-blown glass vessels from which air could be evacuated by vacuum before rarified gases were introduced. The tubes contained positive and negative electrodes, between which a high-voltage current could travel. It was known that gases were poor conductors of electricity due to their neutral charge, but when they were subjected to extremely low pressure and sufficiently high voltages, the electrical discharges they produced could be observed within the tubes, often in lurid colour. It was a procedure that would eventually yield neon lighting.

In preparing equipment for the following day's experimental work, Röntgen darkened his laboratory to gauge the glow that would emanate from a tube when it was receiving electrical input, thereby indicating it was ready for the next series of tests. The tube in question had been encased in a thick, black carton to prevent any exterior light source from interfering with his observations, but in the darkened workspace he noticed a faint glow coming from a paper disc several feet away. That plate had been coated with barium platinocyanide, a known phosphorescent. When Röntgen placed impediments of varying thicknesses between the electrified tube and the disc – even over distances up to two metres – and the fluorescence continued, he understood that the rays being emitted could pass through solid matter.

The most startling revelation came weeks later, when Röntgen enlisted the aid of his wife, Bertha, who placed her left hand on a photographic plate while the tube emitted its rays for fifteen minutes. The result was the now emblematic, skeletal image of her fingers, complete with wedding ring in situ. Bertha Röntgen's response to this ghostly image, produced by emissions of such mystery their discoverer labelled them 'x-rays', was reputedly: 'I have seen my own death.'[10]

The science world was similarly stunned when Röntgen's paper 'On a New Type of Rays' was published in *Nature* magazine in January 1896. The excitement was due not so much to the potential medical benefits of being able to examine the internal structure of animate objects without slicing them open. Rather, physicists saw that discovering a means of measuring these x-rays' interactions with other forms of matter might, in turn, unlock secrets about the structure of atoms.

Months after 'x-ray' became a scientific buzzword, third-generation French physicist Henri Becquerel added a couple more: 'natural radiation'. Becquerel was aiming to establish a link between x-rays and naturally occurring phosphorescence. Having inherited from his father a supply of uranium salts that were known to fluoresce when exposed to sunlight, Becquerel wrapped photographic plates in multiple sheets of black paper and metal foil before exposing them to the salts. He noted that, upon removing the covering, the plates had 'fogged' due to rays that had penetrated the outer layers. By chance, the experiment was repeated on days of thick cloud cover, which showed that the phenomenon – later dubbed 'radioactivity' – was not dependent on the source material's interaction with solar rays.

The notion that naturally occurring elements were capable of releasing measurable forms of matter, which appeared to defy the millennia-old premise that the atoms from which everything was formed were indivisible, would become the central focus of global science for the next half-century. And nowhere would it be more intensely pursued than at the Cavendish Laboratory.

Initially, J.J. Thomson foresaw that if what he described as invisible 'ions' could be released from within these natural elements through the application of electricity, and then examined, they might reveal themselves to be a product of the breakdown of the chemical bonds that held atoms together. That would present an opportunity to venture into a world of matter smaller than had been previously investigated, by studying electricity's passage through gases.

It seems likely that Thomson then encouraged his prized New Zealand research student to embark on that journey with him. It was an opportunity that Rutherford fairly leaped at.

Upon immersing himself in sub-atomic study, Rutherford wrote to Mary Newton: 'The great object is to find the theory of matter before anyone else, for nearly every professor in Europe is now on the warpath.'[11] In a separate missive sent soon afterwards, he warned his fiancée in partially self-deprecating terms: 'I have some very big ideas which I hope to try and these, if successful, would be the making of me. Don't be surprised if you see a cable some morning that yours truly has discovered half a dozen new elements … The possibility is considerable, but the probability rather remote.'[12]

In April 1897, J.J. Thomson became the first to identify a sub-atomic particle. His discovery of the electron – initially labelled the 'corpuscle' but recognised by a new name within months – earned him a Nobel Prize and gilded the Cavendish Laboratory's reputation for world-leading research. It also brought the first revision of the atomic model in almost a century, overturning John Dalton's once revolutionary idea – one that both Thomson and Rutherford had studied in textbooks – that atoms were uniformly solid, unyielding spheres. It had been known as the 'billiard ball' model.

Thomson's radically revised picture that the atom was not an indivisible sphere but comprised electrically charged particles was dubbed the 'plum pudding' theory. He proposed that electrons, represented by pieces of fruit in orderly arrangement, buzzed around in small orbital rings within a gelatinous, dough-like mass of positive charge that countered their polarity.

Thomson's discovery set physics laboratories throughout Europe, and some in North America, into a frenzy of investigation. Their collective quest was to divine more about the 'jolly little beggars',[13] as Rutherford dubbed the charged particles he had been enlisted by his professor to track. That complementary work found Thomson's electrons to be so infinitesimal they were between 2000 and 4000 times smaller than a hydrogen atom.

* * *

Rutherford might have expected that such groundbreaking work alongside Thomson would see his status within Cambridge raised. But while the university's student profile had broadened through its newly inclusive admissions program, other prevailing attitudes at an institution steeped in British rigidity remained staunchly unflinching. In particular, its antiquated financial structure had refused to yield to modernity.

As an entity, Cambridge University was comparatively poor. Its true wealth resided with the network of powerfully autonomous colleges. Some of these – including Corpus Christi, Gonville and Caius, and Trinity Hall – had been founded in the fourteenth century and had accumulated vast assets through investment of their endowment funds. In 2018, it was revealed that the colleges of Oxford and Cambridge universities collectively own more British land – worth £3.5 billion – than does the Church of England.

Consequently, Cambridge in the time of Ernest Rutherford was essentially a company town – its controlling interest being held by the twenty-four constituent colleges, which wielded influence commensurate with their balance sheets.

That power extended to remuneration, and while senior staff, including professors, mostly received inferior salaries, their pay was supplemented by lucrative college fellowships, some worth up to £250 per annum (around £30,000 today), payable for the life of an appointment. This privilege came with apartment accommodation within the relevant college.

A token annual grant was made by the university to the Cavendish Laboratory to cover its running costs, but other expenses – including payment of workshop staff and demonstrators as well as procurement of apparatus – were dependent on the charges levied for teaching and examining. It was a system that instilled in Rutherford the parsimony that would characterise his subsequent tenure as laboratory director.

Rutherford's standing as a geographical and cultural outsider meant that, despite his undisputed talent, his professional future at the Cavendish was bleak. He belonged to a cohort not seen at Cambridge until 1895 – a student who had earned his undergraduate degree elsewhere. And, despite his stellar research output and close rapport with J.J. Thomson, it had become apparent that he would not be considered for a college fellowship until he had served at his adopted institution for at least four years.

Rutherford shared his frustration at that bureaucratic encumbrance from his colleagues in a letter posted to Mary Newton's Christchurch home in June 1898. 'I think it would be much better for me to leave Cambridge, on account of the prejudice of the place,' he wrote, clearly disillusioned at his inferior status. 'I know perfectly well that if I had gone through the regular Cambridge course, and done a third of the work I have done, I would have got a Fellowship bang off.'[14]

Less than three years after landing the scholarship he had believed would assure his future, Rutherford was once again considering his options. And for the first time in his academic life, an alternative opportunity came looking for *him*.

By the final years of the nineteenth century, the Cavendish Laboratory had become not just the proving ground for aspiring physicists, but also the first point of inquiry for rival institutions looking to enter the race for sub-atomic secrets. So when Montreal's McGill University – just turned seventy-five and recently in receipt of a huge bequest from tobacco baron William Macdonald to fund a lavish new physics building – was looking for a suitable figurehead, it was to J.J. Thomson they wrote for recommendations. Thomson then proffered his ideal candidate.

'I have never had a student with more enthusiasm or ability for original research than Mr Rutherford,' Thomson wrote back, 'and I feel sure that if he were elected to the Professorship for which he is a candidate, he would establish a distinguished school of Physics at Montreal.'[15]

The answer was not so straightforward for Ernest Rutherford, however. As much as he had found frustration at the impasses that Cambridge presented, he was mindful of reputation and prestige. In addition, having spent years clawing his way up New Zealand's educational hierarchy to seize the few opportunities on offer, the prospect of renouncing the Cavendish in favour of a colonial posting seemed potentially profligate.

When Rutherford first became aware of the impending vacancy at McGill, replacing another Cambridge alumnus, Hugh Callendar, as Professor of Experimental Physics, he was convinced that Thomson would not support him in pursuing it. And he held doubts about his own suitability for such a senior position, almost as if to dissuade himself from contemplating a shift. As he wrote to Mary Newton in April, 1898:

'I think it is doubtful whether J.J. will want me to go for it. There would probably be big competition for it, all over England, as the average man does not mind going to Canada, though he would bar Australia.'[16]

For once, however, it was a force beyond science – his pledge of marriage to Mary – that would sway him. McGill's departmental head of physics, John Cox, and the university's Scottish principal, William Peterson, travelled to Cambridge to interview their candidate, and confirmed that the position came with an annual salary of £500 (around £60,000 today). That would allow Rutherford not only to repay his outstanding 'grub stake' to his brother, but also to make good on his promise of matrimony. Mary agreed to wait until he had established himself professionally and financially in Montreal, recognising that Ernest's work would always be the third entity in their relationship.

If Rutherford needed further inducement to make the move, the vision that his suitors portrayed of the modern amenities awaiting him in Canada – a sharp contrast to the cramped, already dated conditions at the Cavendish – sealed the deal.

On 5 August 1898, three years after he departed New Zealand and two months before beginning duties in Montreal, a triumphant Rutherford wrote to his future wife: 'Rejoice my dear girl, for matrimony is looming in the distance. I am expected to do a lot of original work and to form a research school in order to knock the shine out of the Yankees.'[17]

Aware that resources in the Empire's far corners could be scarce, before his departure Rutherford arranged for small supplies of the valuable radioactive elements uranium and thorium to be transported the 4800 kilometres to Montreal. This would enable him to resume investigation immediately into the questions that had dominated his final year at Cambridge.

What he had *not* planned was the collaboration he would form with an eager young assistant, who introduced himself by harshly denouncing the very theories upon which Thomson and Rutherford had built their burgeoning renown.

4

'THEY'LL HAVE OUR HEADS OFF'

Canada, 1898 to 1907

Rutherford eyed his junior quarry with a mixture of bristling indignation and mild shock. In light of his experiences with the Dialectic Society in Christchurch, he well understood the adversarial nature of an academic debate. He did not, however, expect to find his work and reputation facing such vehement attack from a boy who had barely started out on his research journey, and had only recently joined McGill's rival chemistry faculty.

Twenty-three-year-old Frederick Soddy's repudiation of Rutherford and Thomson's radical 'corpuscular' theory hung conspiratorially in the air of the Physics Building library as he argued the case in favour of 'Chemical Evidence of the Indivisibility of the Atom'.

At the time Rutherford was sailing for Canada in 1898, Soddy had still been completing his chemistry degree at Oxford University. Despite arriving at Montreal in late 1900, just months before he faced off against the physics professor, he knew of Rutherford's repute from his earlier university studies. He had therefore agreed to lock horns at a debate hosted by the McGill Physical Society in late March 1901, on a subject about which he felt equally passionate. Soddy vehemently rejected the 'plum pudding' theory, which he likened to the ancient alchemists' attempts at transmutation, and was not to be cowed by his formidable foe.

The young chemistry demonstrator took the wind from Rutherford's usually billowing sails by questioning whether electrically charged 'corpuscles' actually constituted matter as chemical science understood it. He also ridiculed the notion that particles could flow from one entity to another in the sub-atomic world as an 'ancient get-rich gimmick foisted upon science'. Rutherford became visibly riled as Soddy continued his attack.

> If as appears the case, radiant matter has lost almost all its ordinary distinguishing properties and appears hardly differentiated from electrical energy, I think the onus of proving that is really a form of matter rests on the new school. And until it is shown that it is affected by gravity, or otherwise possesses features distinct from the ether, there will be no necessity for chemists to modify the Atomic Theory …[1]

Launching into his rebuttal, Rutherford initially struggled to retain civility. The day before, he had written to Thomson at Cambridge, chortling: 'Your corpuscular theory seems to take the field in Physics at present … We are having a great discussion on the subject tomorrow in our local "Physical Society" when we hope to demolish the Chemists.'[2]

All hints of that hubris had now vanished, as Rutherford blustered in righteous defence of his mentor's thesis. He pointed out that 'considerable evidence has been obtained of the production, under various conditions, of bodies which behave as if their mass was only a small fraction of the mass of the chemical atom'.[3]

So intense was the debate that it continued into the following week's meeting. The professor presented his young doubter with a retinue of evidence to substantiate his contention, much of which he had gathered or witnessed at the Cavendish. He also acknowledged the validity of some points Soddy had raised, which helped to soften his initial criticism.

Rutherford's comparative seniority ultimately held sway, and Soddy conceded he might have over-reached. Yet the professor was impressed by his interrogator's inquiring mind, and the conviction with which he presented his views. He began to regard the young man as more an equal than a rival. Soddy, in turn, came to share Rutherford's vision, and was soon enlisted to work in the physics department, helping to track and measure the speed and mass of 'emanations' given off by the radioactive element thorium. It was there that he became a card-carrying devotee.

'I came fully under the influence of his magnetic, energetic and forceful personality,' he would recall of working alongside Rutherford. 'Which, at a later date, was to cast its spell over the whole scientific world.'[4]

*　*　*

McGill might have received its Royal Charter to operate as a university in 1821, but its growth into a diverse, modern educational institution did not begin until decades later, when some of Canada's wealthiest citizens were recruited to bankroll a major building program. The physics edifice to which William Macdonald had put his money and name was of Richardsonian Romanesque influence and had been completed five years before Rutherford's arrival. It featured a distinctive tower that dominated its north-west corner, and the curved portico marking the main entrance was supported by two sturdy columns respectively labelled 'Power' and 'Knowledge'.

While it was similar to the Cavendish, in that it had been purpose-built and therefore contained no iron or steel – even in the heating system – to minimise magnetic interference, it boasted space that Cambridge simply could not spare. Inscribed above the expansive entrance hall's large fireplace was the motto 'Prove All Things', and McGill's new professor of physics quickly surmised that this was a place where he could happily attempt to live that creed.

Rutherford had written approvingly to Mary soon after he completed a torrid sea journey from Liverpool plagued by thick smoke and fog during its final leg along the St Lawrence River.

> I am very pleased with the Physics Building which is very large and fine, six storeys or rather seven, and filled with apparatus. Everything is very bright and polished, in fact almost too much so

for a building where work is to be done ... The Physical Laboratory is one of the best buildings of its kind in the world and has a magnificent supply of apparatus that alone cost £25,000 [around £3 million today].[5]

The modernity of the facilities and fixtures in the new laboratory was undoubtedly an eye-opening improvement on the gloomy, dank workspaces of the Cavendish. Yet the sparsely populated department that Rutherford was to oversee reminded him of his student days in Christchurch. At the time of his arrival at Montreal in late 1898, physics at McGill comprised two professors, of whom he was one, and a scattering of junior instructors and research students.

The resemblance to his homeland extended to Montreal's colonial ambience, and the heavily wooded environs of the city. In warning Mary that the winter weather could be numbingly cold, he assured his wife-to-be that it was a 'very fine place' even amid its bright, frosty chill, and that 'Living, I should imagine, is much the same as in New Zealand.'[6]

However, an obvious difference that Rutherford soon discovered, to his chagrin, was the exorbitant cost of accommodation. He eventually settled on lodgings in McGill College Avenue in downtown Montreal, a short walk from the Physics Building, but the expense thus incurred would carry ramifications for his short-term plans.

In customary fashion, he found solace in a gruelling work regime that meant he often laboured in the McGill laboratory until 11pm or midnight, five nights out of seven. Among his few leisure activities during the warmer weather were cycling forays into the Quebec countryside, occasional rounds of golf and even more irregular games of tennis. His first experience of an arctic winter put paid to those activities, and instead he found distraction by taking the three-kilometre walk across the frozen St Lawrence River, during which he became fascinated by the sight of huge ice floes, and the way they were carved into blocks, to be used as a refrigerant in Canadian households when the spring thaw arrived, or exported to the West Indies.

* * *

For all the accolades and testimonials he took with him to Montreal, Rutherford initially found himself in the shadow of the man he had replaced as Professor of Experimental Physics, Hugh Callendar. His feelings were laid bare in a letter to Mary Newton shortly after his arrival at McGill:

> I am getting rather tired of people telling me what a great man Callendar was, but I always have the sense to agree. As a matter of fact, I don't quite class myself in the same order as Callendar, who was more an engineering type than a physicist, and who took more pride in making a piece of apparatus than in discovering a new scientific truth – but this between ourselves.[7]

It was through the hours and the energy that Rutherford invested in his laboratory work, and the fundamental importance of the discoveries he made, that he soon established himself as not only a talismanic figure for McGill, but one of the world's foremost experimental physicists – with his partner in discovery, Frederick Soddy, closely alongside.

Through his work at Cambridge, examining the properties of Becquerel's mysterious x-rays, Rutherford had already identified two little-understood forces that each betrayed different properties. By employing an electrometer to measure the current of the rays being given off by these forces, Rutherford confirmed they were not homogeneous, and accordingly he assigned them distinctive names. Those that were easily absorbed by aluminium sheets placed over the radioactive source he named 'alpha rays'. The ones that showed greater penetrative capabilities when subjected to similar tests became 'beta rays'.

In later examinations in 1902, Rutherford and Soddy would famously discover that the 'alpha' emanations were ionised helium atoms shot out by naturally occurring radioactive elements, such as radium and the thorium that he had shipped to Montreal. The altogether different 'beta' rays comprised high-speed electrons. A third form of emanations had already been identified, in 1900, by French physicist Paul Villard, and were later named 'gamma' rays, but his discovery of a radiation source that carried no electrical charge and could not be bent by magnetic forces went largely unnoticed at first.

That was likely due to the excitement being created by Rutherford and Soddy, whose work hinted, for the first time, at the extraordinary premise that the disintegration of matter was taking place from *within* the atomic structure, rather than through molecular interaction with outside forces or materials. In line with the beliefs he had espoused at the Physical Society debate, there were times when Soddy was more shaken than spellbound.

One of the pair's most stunning early findings arose from the study of thorium samples conducted in 1902, when they found that the material being emitted showed no evidence of activity – not even when it was bombarded by some of the strongest available laboratory reagents, including platinum, zinc and magnesium, some of which were super-heated. This led them to deduce that what was emanating from thorium – a silvery-black metal that had become the second element (after uranium) to be identified as radioactive – was an inert gas. It was similar, therefore, to the colourless, odourless, non-flammable, non-toxic argon, which had been successfully separated from samples of air by Scottish chemist William Ramsay in 1894. Soddy would later work with Ramsay at University College London.

However, argon was known to be a unique chemical element. So if the conclusions that Rutherford and Soddy reached as they hunched over their ornately precise brass, wooden and hand-blown glass equipment were correct, the process they were witnessing was thorium naturally turning into matter of an altogether different structure. This was the very outcome sought by 'chemists' for centuries – albeit with the aim of transforming otherwise worthless substances into those of infinitely greater value: the very concept against which Soddy had argued so vehemently.

'Rutherford, this is transmutation,' he stammered as the pair stood in the laboratory, transfixed by, and disbelieving of, what they saw. 'The thorium is disintegrating and transmuting itself into argon gas.'

'For Mike's sake Soddy, don't call it transmutation,' Rutherford roared with a laugh, already aware of the controversy this discovery would unleash among their peers. 'They'll have our heads off as alchemists … make it transformation.'[8]

Word of what was being investigated in the physics department began to percolate through McGill, even before it was announced to the wider scientific world, and it left a number of Rutherford's colleagues distinctly uneasy. Such tinkering with the laws of the universe, as they had been set out by Newton, seemed certain to be a source of acute embarrassment for such a young institution beginning to build its global brand.

At a meeting of the McGill Physical Society later in 1902, Rutherford was counselled to cool his heels and delay any publication of his findings in reputable scientific journals. But the professor subsequently found a supporter in the head of department, John Cox, who had interviewed him at Cambridge. Cox suggested that, rather than dishonouring McGill's standing as a serious player in the global physics marketplace, Rutherford's pioneering work on radioactivity might just enhance it. This was, as it turned out, a prudent lesson in the power of self-promotion.

So Rutherford and Soddy ploughed on, working long hours but mindful to avoid the term 'transmutation'. Rather, they substituted 'subatomic chemical change' to denote the process through which a chemical atom spontaneously broke up.

One of the methods Rutherford harnessed to explore changes taking place within an ionisation chamber was to gently fill it with tobacco smoke from the pipe for which he had developed a liking during his separation from Mary. In so doing, he incidentally devised a precursor to the modern household smoke alarm.

* * *

So intransigent was Mary Newton towards the twin vices of drinking and smoking that Rutherford wrestled with how best to advise her he had been seduced by the latter. Barely a year into his Cambridge stint he had penned a heartfelt *mea culpa* to his fiancée.

> A good long time ago, I gave you a promise I would not smoke and I have kept it like a Briton, but I am now seriously considering whether I ought not, for my own sake, to take to tobacco in a mild degree. You know what a restless individual I am, and I believe I am

getting worse. When I come home from researching I can't keep
quiet for a minute, and generally get in a rather nervous state from
pure fidgeting. If I took to smoking occasionally, it would keep me
anchored a bit, and generally make me keep quieter. I don't think
you need be the least bit alarmed ... For I don't think I will ever
become a confirmed smoker.[9]

Smoking had been almost expected of research staff and students at the Cavendish, and Rutherford had acquired a pipe habit despite his earlier pledge. The practice was frowned on at McGill, whose benefactor William Macdonald was an avowed anti-smoking zealot, even though he had accumulated his vast fortune through the sale of tobacco products. Yet the pipe went with Rutherford to Montreal – largely because his fiancée did not.

Soon after arriving in Canada, he had written to Mary advising her that Montreal's steep rents meant he would need some time to build up the cash resources required to bring her to Canada. Consequently, he decreed they would have 'to postpone our partnership for eighteen months from now'.[10]

It was not until June 1900, four years after they had committed to wed, that Ernest and Mary Rutherford finally married in New Zealand. Within a year of settling in Montreal, their only child, a daughter, was born.

Even upon assuming the responsibilities of fatherhood, Rutherford rarely allowed his mind to drift far from his work. After the baby's birth on 30 March 1901, he wrote to his mother at Pungarehu.

The baby, much to Mary's delight, is a she and is apparently
provided with the usual number of limbs. There is much excitement
in college and on the night of her arrival I was toasted at a whist
party. It is suggested that I call her 'Ione' after my respect for ions
in gases. She has good lungs, but I believe uses them comparatively
sparingly compared with most babies. The baby is, of course, a
marvel of intelligence and we think there never was such a fine baby
before.[11]

Rutherford eschewed his colleagues' tongue-in-cheek advice, and the girl was mercifully named Eileen Mary. The family had taken up residence in a modest, but comfortable home in Sainte Famille Street, a short walk from the campus and its Physics Building. In the early years of her married life, Mary Rutherford was a quiet, almost shy soul whose preoccupations in the unfamiliar city were her home, husband and daughter. 'I am not a society woman,' she would confess. 'There's no use trying to make me into one. I love my home and my garden. That's where I belong.'[12]

However, as Ernest's fame began to grow, and his presence was sought at scientific fora and awards ceremonies across North America and Europe, Mary's assuredness grew in step with her worldliness. While the Rutherfords' modest early holidays took them to nearby locations such as Kamouraska, further up the St Lawrence River, by 1903 they were visiting countries such as Switzerland, as Mary began to embrace the trappings of Ernest's new-found celebrity. In late 1904, she and Eileen returned to New Zealand for an extended holiday with family in Christchurch. Her husband finally joined them in May the following year, after completing his first book (entitled *Radio-activity*). It was Rutherford's first visit home since his marriage five years earlier and, as he discovered during a tour of the North Island's thermal districts with his wife and daughter, his profile had increased considerably in that time.

* * *

For Rutherford himself, the rewards lay in a regular stream of vital breakthroughs, many of them commonplace knowledge today. Work increasingly consumed him, and even on Christmas Day of 1901 – the first he had shared with his wife and then nine-month-old daughter – he visited the laboratory to record radioactivity emanations.

By tracking levels of radioactivity emitted by naturally occurring elements, he and Soddy found that the rate at which that output decreased followed a consistent geometric progression. In the case of the thorium compound he tested, its output was halved after sixty seconds, and then halved again a minute later. This process was applied across numerous elements and compounds, and the time it took any given sample to reduce its level of radioactivity by half could be charted on an exponential curve – introducing the notion of radioactive half-life.

That method could then be used to determine the age of any material that emitted detectable rays, such as uranium. By extrapolation, Rutherford was then able to calculate a creation date for the earth itself, of around 4.5 billion years ago, give or take 50 million: roughly the half-life of the predominant naturally occurring uranium isotope (uranium-238). This was a discovery that stunned geologists and biologists, who believed the earth was between 20 and 40 million years old. Many of them viewed Rutherford's 'radiation clock' with deep cynicism, even up until the oldest known mineral exhibit – a sliver of zircon hewn from the inland Jack Hills range north of Perth in Western Australia – was found in 2001, and dated to almost 4.4 billion years old.

Rutherford would later delight in recalling his encounter with McGill's geology professor, Frank Dawson Adams, during a walk across campus in the early years of the twentieth century.

'How old is the earth supposed to be?' Rutherford inquired of his colleague, with mischief sparkling in his blue eyes.

'One hundred million years,' Adams replied, after momentarily pondering the calculations that Lord Kelvin had made four decades earlier, and which had remained widely accepted throughout the science fraternity.

Rutherford then grinned knowingly as he reached into his pocket to extract a lump of pitchblende, a radioactive mineral prized for its high uranium content. 'I know for a fact that this piece of pitchblende is 700 million years old,' he declared before walking on, chuckling, while Adams was left muttering into his slipstream.[13]

The method Rutherford had employed to underpin this assertion was disarmingly straightforward. By determining the proportions of uranium and radium in the pitchblende, then calculating the release of alpha particles – which he had deduced were helium atoms – he could measure the levels of helium that remained within the rock. Then, through the application of simple division, he was able to nominate the period in history in which the substance, in its complete form, had been born. He determined that this period was up to 200 times earlier than science had previously accepted.

He and Soddy had noted the pattern of 'transmutation' in other elements, and believed that the emanations they so diligently tracked were smaller, lighter atoms being spat out by the source material. The energy release that accompanied that process came from the breaking of bonds that held these atoms together. Ernest Rutherford and his occasionally anxious young accomplice had floated a concept that the world would come to recognise, albeit decades later, as atomic energy.

In the years that followed, Rutherford repeatedly scoffed at claims that the breaking of atomic bonds might unleash catastrophe, by pointing out that radioactivity was as old as the earth itself: a birthdate that he had calculated, and which had passed without leaving any evidence of an atomic explosion.

But he could not resist slipping in his favourite jokes when discussions about the impact of sub-atomic research turned to the potential perils it held.

'Could a proper detonator be found, it's just conceivable that a wave of atomic disintegration might be started through matter which would, indeed, make this old world vanish in smoke' was one attributed to him in an article written by a Cambridge associate.[14]

Another was pithier, and more prescient. 'Some fool in a laboratory might blow up the universe unawares,' he would occasionally warn, before letting loose a volley of booming laughter.[15]

Rutherford's public refutations of the perils of probing the power that held together atoms, and the dark humour he liked to employ when undertaking that work, helped to mask his own innate concerns as to the forces he and others were tapping. As his investigations took him deeper into the heart of matter, and his appreciation of the energies residing there became clearer, he saw the potential of that power if it could somehow be harnessed. However, he and Soddy also feared for the planet if those discoveries were made against a backdrop of, and for the purpose of prosecuting, international conflict.

After their research partnership dissolved in 1903, with Soddy returning to England to take up the position at University College

London, it seemed the younger man had also gleaned Rutherford's freakish talent for predicting the future. The following year, Soddy delivered a lecture on radium to the Corps of Royal Engineers in which he articulated his thoughts on the potential application of the work he and Rutherford had pursued.

> It is probable that all heavy matter possesses – latent and bound up with the structure of the atom – a similar quantity of energy to that possessed by radium. If it could be tapped and controlled what an agent it would be in shaping the world's destiny! The man who puts his hand on the lever by which a parsimonious nature regulates so jealously the output of this store of energy would possess a weapon by which he could destroy the earth if he chose.[16]

Soddy's career would later take him to the Universities of Glasgow and Aberdeen, then back to Oxford and – in 1921 – garner him a Nobel Prize for his work on radioactive decay and formulating the concept of isotopes. He was in no doubt as to who had been the most direct contributor to his success.

Upon learning of his Nobel triumph, he wrote to Rutherford, 'acknowledging the debt I owe you for the initiation into the subject of radioactivity in the old Montreal days. But for that, I suppose the chance of my ever getting the Nobel would have been exceedingly remote.'[17]

* * *

During his nine years at McGill, Rutherford would be involved in drafting an extraordinary sixty-nine scientific papers: more than the total output of his subsequent research career. Yet it was the substance more than the volume of his work that saw him shine ever brighter in what was the lustrous firmament of worldwide physicists.

In 1903, Rutherford was elevated to fellowship of the prestigious Royal Society in London and, on travelling to Britain to receive the honour, he grasped the opportunity to introduce Mary – accompanied by her mother – to his growing group of friends and contemporaries in Europe's science community. Among them was Ernest's former Cavendish Laboratory neighbour, Paul Langevin, who had overcome his initial wariness of Rutherford as a 'force of nature' at Cambridge. In 1902, Langevin had returned to France to study with Pierre Curie.

The Rutherfords arrived in Paris on the same June day that Curie's thirty-six-year-old wife, Marie Sklodowska Curie, received her doctorate from the Sorbonne, to the undisguised resentment of the otherwise exclusively male French scientific elite. Langevin invited both couples for dinner, and Rutherford – whose university experience in New Zealand was of male and female students accepted and recognised equally – struck up an immediate rapport with the intense, quietly spoken Madame Curie, who was dressed head to toe in black.

As dinner service finished and midnight approached, the quintet retired to Langevin's garden. Pierre Curie then produced a small tube containing a trace sample of the preciously rare radium that he and his wife had painstakingly separated from pitchblende in their laboratory. The vessel, partly coated with zinc sulfide, glowed ghostly in the darkness as the radioactivity worked its magic. But even under the soft light of a Parisian moon and the eerie luminescence from the glass flask, Rutherford noticed the raw and inflamed state of Professor Curie's hands, which at times had difficulty gripping the receptacle.

Rutherford immediately thought of his own laboratory, where he had noticed that items on workbenches, even their notebooks, had begun to return low-level radioactivity readings. Even though he had fortuitously suffered no ill effects from his pioneering exploration of the embryonic science of radioactivity, warnings as to the lethality these substances might possess were starkly, if silently, conveyed by Pierre Curie's gnarled, trembling fingers.

* * *

The increase in Rutherford's profile brought commensurate demand for his person. The year after gaining his Royal Society fellowship he was awarded its Bakerian Medal, which recognised outstanding achievement in 'natural history or experimental philosophy'. Recipients of the prestigious award were also required to deliver a lecture, a tradition first practised in 1775, and in 1904 Rutherford spoke at length on the theory of radioactivity.

Among previous winners of the Bakerian Medal were some of the great names of British science – Humphry Davy (who in the early 1800s discovered, through electrolysis, elements including sodium, potassium, magnesium and calcium), Michael Faraday, James Clerk Maxwell and J.J. Thomson, as well as German-born physicist Arthur Schuster, who had worked alongside Maxwell at the Cavendish.

In 1907, Schuster announced he would stand down from his position as Langworthy Professor of Physics at Manchester University – essentially the same post as had once been held by Balfour Stewart, whose *Physics* primer had proven so influential – if Rutherford would agree to succeed him. Rutherford had already rejected multiple offers from American institutions of growing repute, including Columbia and Stanford. As he continued his work at McGill, he was nominated for the directorship of Washington's Smithsonian Institution, and three times within five years he was approached to take over the physics chair at Yale University.

However, his regular visits to Britain and Europe to receive awards and acclaim had stirred his yearning to be closer to science's historic heart, and the tabling of a salary that was double that of McGill made the decision clear. Despite having purchased a parcel of land in Montreal, on a hillside that offered uninterrupted views of the Lake of Two Mountains, Ernest and Mary shelved plans to build a house and forge a life in Canada.

Instead, Rutherford accepted the position on offer at the University of Manchester. So significant was this recruiting coup considered that Britain's parochial press shrilled with the news that, after almost a decade at McGill, New Zealand's most celebrated expatriate was coming 'home'.

The world would be forever changed as a consequence.

5

THE ATOM SMASHER

Manchester, 1907 to 1919

'By thunder,' Rutherford roared, slamming his meaty fist down on the speaker's lectern.

He was addressing his first meeting of the University of Manchester's combined science faculties. The newly appointed Langworthy Professor had just learned that – in the course of the horse-trading that had prised him from Montreal – floor space, equipment and personnel originally earmarked for his physics department had been appropriated by rival chemistry professor Harold Baily Dixon.

It wasn't only the thrift instilled within him by an early life on the land and the frugality he had experienced at the Cavendish that made Rutherford proprietorial towards the tools of his trade. In the weeks preceding his departure from Montreal in May 1907, McGill had lost two major buildings – engineering and medical – to fires thought deliberately lit. He was therefore not about to surrender crucial resources required for his ongoing experimental work to a force as readily extinguishable as a fellow don.

If the shocked silence did not reassure him his pre-emptive salvo had struck its mark, then his second attack surely did. He followed the target of his ire out of the meeting, shouting theatrically down the corridor: 'you are the fag end of a bad dream'.[1]

As was now wincingly apparent, Rutherford had arrived at

Manchester with a bang. Within weeks of the meeting, he had regained all that had been purloined and, just as symbolically, he had announced to the Science Faculty – and the entire university – that he was indeed a force of nature. While his demeanour in polite society was affable and unaffected, he could unleash an explosive temper when events refused to bend to his will, or careful plans were laid waste.

If reminded of such intemperate outbursts, Rutherford was known to wear his directness as a badge of honour. As the principal guest at the Cavendish's annual research dinner in 1911, he would laugh uproariously when the meeting's chair introduced him by noting that, among all the budding young physicists to have graced the famous laboratory, Rutherford was peerless when it came to swearing at equipment.

Rutherford was not, however, the first scientist of such eminence to grace Manchester's echoing corridors and damp courtyards. The institution had begun its life more than eighty years earlier as the industrial city's Mechanics Institute, and had counted among its founders the acclaimed chemist John Dalton, author of the first atomic model. In 1880, it had received a royal charter to become England's first 'civic university' (where students from all religions and backgrounds were accepted on an equal basis), and duly adopted the formal title 'The Victoria University of Manchester'.

Upon arriving in Manchester in late May, the Rutherfords spent a week or two searching for a house before opting to take summer holidays on the Devonshire and Cornish coast. When they returned north for the beginning of the new academic year, they moved into a comfortable home in Withington, about three kilometres from the university. Withington had once been a standalone village, but so rapid had the city's growth been during the Industrial Revolution and beyond that it had now been subsumed into Greater Manchester.

With its hum of machinery and ever-present blanket of factory smoke or low cloud – more often both in tandem – the city made for a rude reminder to Rutherford of his previous experiences in England. But it was a far greater culture shock for Mary and Eileen. Both had come to enjoy the clean air and natural attractions of Montreal and the Quebec landscape, so adapting to grim suburban life brought its challenges.

In the years before Rutherford's recruitment, Manchester's physics operation had become recognised as the second most influential in Britain after the Cavendish. That had been mainly due to Arthur Schuster. He and his family had moved to Manchester from their native Frankfurt in 1870, and the Schusters formed part of an enduring connection between the university and Germany. That included campus architecture, as well as research-led teaching philosophies. When Rutherford arrived in 1907 there was also a distinctly Germanic influence within his laboratory team.

One of those men was technician Hans Geiger, who would play a pivotal role in Rutherford's history-defining discoveries. Working alongside Rutherford, Geiger also invented a revolutionary radiation detection and measuring device that would famously bear his name. Rutherford already held a high regard for German science and its practitioners, thanks to his collaborations with his McGill research student Otto Hahn, who would later emerge as a central figure in the evolution of nuclear physics.

Although Manchester was not the beneficiary of the sort of largesse William Macdonald had showered upon McGill, Rutherford was pleased to find his facilities more than adequate, once his pilfered resources had been recovered. 'The laboratory is very good, although not built so regardless of expense as the laboratory in Montreal,' he wrote to his mother in New Zealand at June's end in 1907.[2]

The initial impression that many at Manchester gained of Rutherford, apart from his capacity for volcanic outbursts, was that he seemed much younger than his thirty-five years. In another letter to his mother, Rutherford delighted in recounting a story from very early in his tenure, when Schuster was still at the university and playing host to Japan's minister of education, Baron Kikuchi. After Rutherford had been introduced, the distinguished visitor said discreetly to Schuster: 'I suppose the Rutherford you introduced me to is a son of the celebrated Professor Rutherford'.[3]

Having announced his arrival so emphatically, Rutherford also resumed his examinations of radioactivity with vehemence, and rapidly made startling progress. First he had to overcome a critical shortage of radioactive material from which he could source the alpha particles that were essential to his research. However, once he had won the administrative battle to secure supplies of the precious radium, he began writing more scientific history.

Initially he sought to achieve that success by confirming his theory that particles being given off by the radium were of a different atomic structure from their source material – thereby proving the process of transmutation that had so alarmed Soddy.

An individual atom of matter had never before been detected, and doing so would take some experimental foresight. It was known that when an electrical charge was passed through alpha radiation particles that had been isolated within a vacuum tube, the bands of colour that were produced could be studied through a spectroscope. The fact that individual elements were known to produce telltale spectral lines in the form of coloured bands separated by solid dark blocks – each as unique as individual fingerprints – meant Rutherford was able to prove the radioactive emanations in the tube were positively charged helium atoms. He and Soddy had theoretically established as much at McGill years earlier, but this represented the first time a single atom had been identified. It also showed definitively that radioactivity released matter of a different atomic structure.

* * *

The lofty place that Ernest Rutherford had come to command within the fast-changing world of sub-atomic exploration was formalised in 1908, the year after he left Canada, when he became a Nobel laureate – even if he won the honour in the field of chemistry, 'for his investigations into the disintegration of the elements, and the chemistry of radioactive substances' undertaken at McGill.

In accepting his prize from the King of Sweden on 11 December 1908, Rutherford was later said by some gathered at Stockholm's Musical Academy to have looked 'ridiculously young' and also to have delivered the evening's best speech.[4] In his address, Rutherford joked that he had seen many transformations in laboratories, but that the quickest one was the Nobel committee's 'instantaneous transmutation' of him from physicist to chemist.[5] Once the laughter had subsided, he proceeded to astonish his audience by detailing his most recent finding at Manchester: that particles expelled by the naturally occurring element radium were shown to be atoms of an altogether different element, helium.

In light of the spectre that nuclear science would come to cast upon the world, there was perhaps an irony in this recognition of Rutherford's work. After all, Alfred Nobel – whose name also adorns the accolade universally associated with peace – built the fortune that perpetually funds his eponymous prizes through his invention of dynamite. It was an innovation so effective that Nobel vainly believed – like the architects of the atomic bomb – it might put an end to all war.

Ernest and Mary arrived back in Manchester just days before Christmas, and while eight-year-old Eileen was anxious to see the festive gifts her parents had brought back from the Continent, a much larger windfall loomed. The Nobel Committee for Chemistry's cash prize of almost £7000 (worth around £800,000 today) meant that, for the first time in his already celebrated life, Rutherford was a man of means. He immediately wired sizeable gifts to his parents, brothers and sisters in New Zealand, and while a bulk of the remaining purse was prudently invested, he also splashed on a new Wolseley Siddeley motor vehicle.

Rutherford had first witnessed the marvel of the 'horseless carriage', as it was known, while at the Cavendish in 1897, during a visit to the Crystal Palace – the relocated glass cathedral that hosted the 1851 exhibition that indirectly had determined his future. At the time, he wrote: 'I was not very much impressed with the machines as vehicles, but I expect they will come into very general use shortly. They say the expense of running about 12 miles an hour [nineteen kilometres an hour] is about one penny per hour, which is rather cheaper than a horse.'[6] Over the intervening decade, however, Rutherford had developed a keen interest in motorised vehicles and the autonomy they could provide him and his family.

The fourteen-horsepower, four-seater tourer was described by Rutherford as ideally suiting their need for 'quiet travelling', and as 'a means of getting fresh air rapidly' by escaping Manchester's choking pollution. However, as he explained to his mother in April 1910, he professed little experience behind the steering wheel of such a vehicle.

We got our car on Good Friday and spent three days running around while I practised driving. We started on our tour Tuesday, and have so far gone 500 miles [800 kilometres] … We have enjoyed ourselves very thoroughly. I have learnt to drive fairly well without a single incident, even of running over a chicken.

    A car is very easy to manage and far more under control than a horse. We average about 17 miles an hour [twenty-seven kilometres an hour] over country, and on a good road run along freely at 25. We can do 35 or 40 if we want to, but I am not too keen on high speeds with motor traps along the road and a ten guinea fine if I am caught. These are the woes of motorists that I hope to avoid![7]

The spoils of Rutherford's new-found means were not channelled solely into meeting his automotive fancies. Mary welcomed the installation of a small heating unit in their Withington house that warmed the ground-floor rooms, a luxury that also meant Eileen was able to invite schoolfriends for a birthday party, even though Manchester remained in the grip of lingering winter when she celebrated the occasion in late March.

Mary also took on the role of regular social entertainer, whether that was at the Saturday-night suppers she hosted in the home's white-walled dining room before the male guests adjourned to Ernest's study, or the Sunday-afternoon teas in the drawing room that preceded a quick run across surrounding countryside in his beloved motor car.

Not that Rutherford was about to throttle back on his laboratory work by assuming the satisfied semi-retirement of a comfortable academic. For a year or more, he had been applying his formidable intellect to the next great sub-atomic mystery.

* * *

Geiger later retained vivid memories of the scene that would greet him when he went in search of Rutherford to begin another session of counting alpha particles.

I see his quiet research room at the top of the physics building, under the roof, where his radium was kept and in which much well-known work on the emanation was carried out. But I also see the gloomy cellar in which he fitted up his delicate apparatus for the study of the alpha rays. Rutherford loved this room. One went down two steps and then heard from the darkness Rutherford's voice reminding one that a hot-pipe crossed the room at head level, and to step over two water pipes. Then finally, in the feeble light, one saw the great man himself seated at his apparatus and straightaway he would recount in his own inimitable way the progress of his experiments, and point out the difficulties that he had overcome …[8]

Rutherford's mission, during these intensive and reclusive sessions, was to establish the speed and properties of those radioactive particles being emitted from precious supplies of radium. The element was sparingly used in experiments, in the form of radon gas that could be 'milked' every few days from the mother lode.

Each gram of radium, the most bountiful source of Rutherford's alpha rays, was believed to release these radioactive particles at a rate of around 34 billion per second. His ambition was to unleash them, directed by a magnetic field, through a narrow slit from within a lead-lined box and train them at a target – as if he were firing them from a gun.

The stream of alpha particle 'bullets' would then be aimed at a small glass plate coated with zinc sulfide, the same material Pierre Curie had applied to his glass tube, which would glow when exposed to radioactivity. Given that it was impossible to see sub-atomic particles travelling at one-seventh the speed of light with the naked eye, Rutherford would sit hunched over a microscope placed behind the plate and manually record each 'scintillation', as the phosphorescent pinpricks of an alpha particle striking the treated surface were known.

During experiments he had conducted at McGill, Rutherford had noticed that some particles could be deflected from their direct path by a degree or two if forced to pass through solid matter, such as delicate mica sheets – silicate mineral, around three-thousandths of a centimetre thick. This suggested to him that some particles were being bumped off track.

When the experiment was repeated using thin strips of foil fashioned from various metals, the degrees of scattering became more and more pronounced as the atomic weight of the metals increased.

This method required the experimenter to sit in the darkened basement for half an hour until their eyes adjusted to the gloom and they could begin to count the fleeting scintillations. As a result, the useful viewing time for each individual was restricted to around thirty minutes.

Rutherford decided it was close work best suited to young undergraduate student Ernest Marsden. The twenty-year-old, along with Geiger, began a series of experiments using gold foil six-one hundred thousandths of a centimetre thick – roughly the equivalent of 400 gold atoms – placed at an angle of forty-five degrees to the radioactive bullets.

Most of those fired at the foil passed directly through, while a number deviated by a degree or two, as per previous findings. But there was also evidence that some alpha particles rebounded from the foil at an angle of ninety degrees.

Intrigued, Rutherford then instructed that the target plate be positioned virtually alongside the particle 'gun'. The men were stunned when scintillation marks showed that around one in 8000 particles bounced back at an angle greater than ninety degrees, while some scintillations were even detected at 150 degrees or more. In other words, these bullets flying at vast speeds were occasionally being flung back in the direction from which they came by a sheet of matter so thin that most other radioactive atoms sailed through it, unhindered.

'It was quite the most incredible event that has ever happened to me in my life,' Rutherford memorably said of his most famous experiment. 'It was almost as if you fired a 15-inch shell at a piece of tissue paper, and it came back and hit you.'[9]

Neither he nor Marsden could understand how this result was possible. Initially, Rutherford reasoned that the positively charged helium bullets had been repelled by an electric or magnetic force, but then he estimated it would take the equivalent of millions of volts to force a fast-moving atom of that size and substance into reverse. So he engaged his favourite piece of laboratory equipment: his capacity to sit quietly and think.

For more than a year, he ruminated over the results while continuing other experiments and overseeing those of his students, giving lectures and writing papers.

It was a neat symmetry that he finalised his working theory while tucking into plum pudding on Christmas Day 1910. He was now convinced he could no longer subscribe to J.J. Thomson's theory, which likened atomic structure to that same dessert. Rather, his new vision was built upon the celestial bodies swirling in the night sky.

One morning early in 1911, he appeared in the doorway of Geiger's office bearing an enormous grin. 'Rutherford, obviously in the best of spirits, came into my room and told me that he now knew what the atom looked like and how to explain the large deflections of the alpha particles,' Geiger would recall.

Rutherford then sat down, and became the first human to explain the atom in accurate detail.

In essence, Rutherford's conception of the atom – which was little different from the way it is understood today – was that it mostly comprised empty space. He theorised that at its core (he did not begin using the term 'nucleus' until several years later) was a dense, hugely powerful particle that was responsible for deflecting the alpha bullets through electrostatic repulsion. A few much smaller and lighter pieces of matter orbited this highly charged core in concentric rings.

The comparative scale of the model prompted the oft-cited analogy that, if the atom were expanded to the size of St Paul's Cathedral in London, the particle at its heart would appear as roughly the size of a single fly. However, the density of that tightly packed core would be so great it would account for 99.975 per cent of the structure's total mass.

The reason why most of the alpha particle bullets fired at the gold foil sailed through unhindered was that the chances of directly hitting the central core of each atom was remote. But when that happened, the bullets were repulsed at a stunning velocity.

Rutherford's radical revision of the atom did more than make J.J. Thomson's plum pudding model redundant. 'The Rutherford model of the atom was like a solar system,' Mark Oliphant liked to summarise years later. 'The nucleus taking the place of the sun, and the electrons that of the planets.'[10]

It fundamentally changed science's understanding of the essential blocks from which the world is built. No longer was the atom the solid billiard-ball representation conceived by Dalton, nor gelatinous dough, as Thomson had subsequently surmised. It was now revealed, by the man whose curiosity had been stirred in the rural backwaters of remote New Zealand, that the entity that makes up all known substances and being is essentially made of nothing. It is no more than a tiny, super-charged pinprick at the middle of a whirring, blurring void.

From what would become neatly known as the 'gold foil experiments', an entirely new research field – nuclear physics – was born. Arthur Eddington, the astronomer who had worked in the Cavendish Laboratory and would famously go on to help prove a number of Einstein's theories, proclaimed Rutherford's discovery the most important development in science since Greek philosopher Democritus proposed the existence of the atom in the fifth century BC.[11]

Baron Bowden, an influential scientist and educator who was examined for his PhD by Rutherford and Oliphant in tandem, was unequivocal as to the impact of this moment on subsequent history.

> The achievements of Rutherford, I am sure, would not have been rivalled by anyone else for at least ten to fifteen years. If nuclear physics had not got started then, and had gone on at the usual sort of rate, we should never have arrived at the possibility of the nuclear bomb at the beginning of the Second World War.[12]

* * *

Sceptical scholars soon wondered why the negatively charged electrons that supposedly hurtled around the positively charged core of Rutherford's new atomic model did not follow Newton's laws. According to those centuries-old principles, they should lose their energy by radiating it away – at which point, they would be drawn into the super-charged core through electrical attraction, thus rendering the entire structure unstable.

To explain this anomaly, Rutherford had to enlist the expertise of a theoretical physicist. An avowed experimentalist, he made no secret of the suspicion he held for those whose research skills were best exhibited on a blackboard rather than a workbench. He was known to mock-seriously chide those of his students and staff who dared discuss matters of theory by warning them: 'don't let me catch anyone talking about the universe in my department'.[13]

However, among the Manchester 'boys' was a theoretician who – as would later happen with Mark Oliphant – had been drawn into Rutherford's orbit by the force of the professor's vision. During the 1911 Cavendish Laboratory dinner at which Rutherford was lovingly reproached for his impatience towards apparatus, Niels Bohr became so inspired by the irrepressible New Zealander's speech that he transferred to Manchester University soon afterwards.

Tall and charismatic, twenty-seven-year-old Bohr was a gifted sportsman who had played top-level soccer in his native Denmark alongside his brother, Harald, who had gone on to represent their country at the 1908 Olympic Games. Bohr became not only a favourite of Rutherford but also of the professor's wife, Mary, who adopted the Dane as a surrogate son. If anyone challenged Rutherford about the incongruity of his closeness to a theoretician, the former Nelson College rugby forward would opine, 'no, no, Bohr's different – he's a football player'.[14]

Bohr applied the quantum theories that had been initially developed in 1900 by German Max Planck to solve the quandary of Rutherford's atomic model. The basis of Planck's assertions, which would revolutionise the understanding of energy forces, was that light and other electromagnetic waves were emitted in discrete packets of energy that he dubbed quanta. In 1912, Bohr essentially showed that when electrons came into direct contact with energy forms such as light or radioactive heat, they would absorb that energy and jump into an altogether new orbit path, which would prevent them from collapsing into the atom's nucleus. A quantum leap, as it would later be called.

Consequently, by 1913, the world would know the end product of years of detective work as the Rutherford–Bohr atomic model.

* * *

By this time, Rutherford was without peer in experimental physics. While his laboratory time at Manchester was largely devoted to counting alpha particle scintillations and further testing his ideas on atomic structure, he also maintained a hectic schedule of scientific lectures, international conferences and exclusive social events. These included the 1912 function at Windsor Castle to celebrate the Royal Society's 250th anniversary, where he was introduced to King George V and Queen Mary. The following year, at the 1913 meeting of the British Association for the Advancement of Science held in Birmingham, he helped organise for Marie Curie to travel to Britain to receive an honorary degree from Birmingham University. She responded to the gesture by noting, in a rare media interaction, that Rutherford was 'the one man living who promises to confer some inestimable boon on mankind as a result of the discovery of radium'.[15]

When the 1914 New Year Honours list was published, the forty-two-year-old son of a New Zealand flax farmer and his schoolteacher wife was confirmed as Sir Ernest Rutherford. He responded to the cascade of well-wishers with typical humility, writing to his former McGill research student Otto Hahn that he did not place great significance upon 'such forms of decoration for they have obvious disadvantages in the case of a scientific man like myself'.[16]

He gained greater pleasure in documenting thirteen-year-old Eileen's excitement at the succession of telegrams that greeted the news – though she voiced adolescent concern that neither her father nor 'Lady Rutherford' (as Mary had become) carried sufficient 'swank' to justify such recognition. Upon seeing her father clad in the outfit he would wear to Buckingham Palace to be knighted, complete with ceremonial sword dangling at his side, she teased that he resembled a 'somewhat superior footman'. Accordingly, Rutherford self-mockingly advised her that 'henceforth, young lady, you may address me as "Sir Ernest"'.

If Rutherford was not especially impressed by his new societal status, he could appreciate the benefits it afforded. Later that year, he was prominent among 300 or so delegates who sailed en masse to

Australia for a series of British Association for the Advancement of Science meetings. This event had been organised primarily to meet an overwhelming curiosity among members to study the continent's egg-laying monotreme, the platypus.

As Ernest and Mary prepared to sail for the southern hemisphere in mid-1914, he expressed his distress over the assassination of Archduke Franz Ferdinand in Sarajevo. 'The Hapsburghs [sic] have a very tragic family history,' he noted.[17]

That tragedy was about to be visited upon the world in the form of brutality that not even the prescient professor could have foreseen. It was while the flotilla of luminaries made their way from Perth to Adelaide, where the first four-day congress was scheduled to begin on 8 August, that Britain declared war on Germany.

* * *

The news broken by the headmaster to Mark Oliphant's high school class had a more profound impact on Britain's science fraternity as they disembarked at Port Adelaide. The RMS *Orvieto*, on which many of the association members had travelled, was immediately requisitioned for war duties, and carried the first wave of Australian troops to the distant battlefront.

Rutherford, however, continued with his lecture commitments as planned, and also went ahead with the New Zealand visit he and Mary had appended to the journey. In Christchurch, Rutherford was afforded a civic reception and delivered a lecture at Canterbury College entitled 'Evolution of the Elements'. It was the same topic on which he had made his first presentation to the college's scientific society before leaving for Cambridge almost twenty years earlier. He and Mary also returned to their families for the first time since he had earned his Nobel Prize and knighthood. They then undertook a leisurely return to England via Canada, arriving home in Manchester soon after New Year 1915.

However, other members of the British Association delegation had made a beeline for Britain as soon as war was declared, to fulfil their patriotic duty. Among them was Henry Moseley, another of Rutherford's

Manchester 'boys', who was considered among the most brilliantly gifted. Moseley, known to all as Harry, had gained kudos at an early age for establishing the sequence of atomic numbers that re-ordered Dmitri Mendeleev's 1869 table of elements. Like Soddy, Marsden and later Oliphant, Moseley had become a revered member of Rutherford's student coterie because of his rare scientific insight, and appetite for hard work.

Moseley had enlisted as a signalling officer with the 38th Brigade, which made landfall at Gallipoli on 6 August 1915. He was unaware of Rutherford's entreaties to officials to find the young man means to contribute to the war effort through scientific work, and thereby have him recalled from active service. In the course of telephoning through an order four days after deploying on the Gallipoli Peninsula, Moseley was shot through the head by a sniper and died aged twenty-seven. As per his will, his earthly wealth was then bequeathed to the Royal Society to be used for the furthering of scientific research. Rutherford wrote a moving obituary for *Nature* magazine.

> It is a national tragedy that our military organisation at the start was so inelastic as to be unable, with a few exceptions, to utilise the offers of services of our scientific men except as combatants in the firing line. Our regret for the untimely death of Moseley is all the more poignant because we recognise that his services would have been far more useful to this country in numerous fields of scientific enquiry rendered necessary by the war than by the exposure to the chances of a Turkish bullet.[18]

Like almost everyone who lived through the hardships and horrors of the First World War, Rutherford came to harbour myriad personal reasons to abhor modern warfare and its increasingly sophisticated weaponry. His uncle – Martha Rutherford's brother, Archie Thompson – sailed from the southern dominions to answer the call but did not survive the Gallipoli campaign. Rutherford's future son-in-law and Cavendish colleague, Ralph Fowler, was severely wounded on the same stretch of peninsula and would spend the rest of his life tortured by serious lung ailments.

The youngest son of William Bragg, Rutherford's early inspiration and lifelong friend, died at Gallipoli after both his legs were blown off by an artillery shell.

Rutherford, too, made a concerted contribution to the war effort – though in keeping with his views, it came not through military service but scientific expertise. The improvements he helped bring to the field of underwater warfare, many of which were trialled in the water tank he had built in his basement workroom at Manchester University, yielded the technology that became recognised as sonar.

Often, however, scientific input into areas of military expertise was conceitedly opposed by the uniformed class. Rutherford repeatedly butted heads with infuriatingly intractable military men whom he saw as suspicious and resentful of civilian scientists. On one occasion, he was sent to Hawkcraig base near the mouth of Scotland's Firth of Forth to investigate the use of underwater microphones in submarines – only to be abruptly told he would not be granted access to any of the navy's fleet because they were all required for active duty.

It was understandable, therefore, that he returned to his laboratory whenever possible to exercise his preference for sub-atomic matter over matters submarine. Near the end of the conflict, he missed most of a meeting held by the Anti-Submarine Committee, of which he was a key member. As he explained in a message sent to unimpressed delegates, he needed to complete a round of experiments in his laboratory, where he believed he had created history by successfully splitting the atomic nucleus. 'If this were true,' his apology offered, 'its ultimate importance is far greater than that of the war.'[19]

Rutherford also enjoyed poking private fun at bumptious individuals who exuded self-importance but no self-awareness. 'He's like the Euclidian Point,' he said of a man who fitted that assessment, drawing deeply on his pipe for dramatic effect. 'He has position without magnitude.'[20]

* * *

Rutherford's Manchester colleague Ernest Marsden, who had been integral to the gold foil experiments, was another to choose a direct role in the war effort. Marsden had already accepted the position of Professor of Physics at New Zealand's Victoria University when war erupted, having been recommended for the job by Rutherford. But soon after arriving in Wellington to take up his new role, he signed up with the New Zealand Expeditionary Force, and his contribution as a signals engineer in Europe earned him a Military Cross.

It was while Marsden was undergoing military training that Rutherford inquired whether his former student would object to him revisiting some of the research work begun by Marsden at Manchester before the outbreak of fighting. On receiving his former student's blessing, Rutherford ran his own series of tests using alpha particles to bombard atoms of hydrogen. He was intrigued by Marsden's reports that some of the emanations had slammed into the scintillation screen with far greater force than that used to originally propel the alpha bullets.

At war's end, before he sailed home to New Zealand, Marsden made a farewell visit to Manchester University, where Rutherford hurried him to the laboratory to take a look at what his continuation of the earlier experiments had revealed.

Marsden saw that Rutherford had replaced his original glass apparatus with an elegant brass cylinder that was less susceptible to radioactive contamination. He explained how he had filled it with dry air, then with water, then oxygen and finally carbon dioxide as he searched for a source of the powerful hydrogen atoms that were being detected.

None of these substances produced startling results on the scintillation screen when hit with a barrage of alpha particles. But when the chamber was filled with pure nitrogen, the screen lit up like the Western Front in miniature.

Rutherford then explained to Marsden how he had used magnetic fields to verify that the particles hitting the screen were positively charged hydrogen atoms that had been 'chipped off' the nitrogen nuclei. The process had effectively changed atoms of nitrogen into altogether different elements: hydrogen and oxygen.

It was the same form of transmutation through natural radioactivity that Rutherford had witnessed in his laboratory at McGill. Only this time it had been manufactured artificially in his darkened Manchester basement.

As he sat with colleagues waiting for their eyes to adjust to the blackness ahead of the never-before-witnessed spectacle, Rutherford warned with a laugh: 'you know, we might go up through the roof'.[21]

* * *

Rutherford was in France, on Admiralty business related to submarine research, in November 1918 as the war entered its final days. He noted the contrast between the high spirits of Parisians and the sombre sight of rows of captured German heavy artillery lined up along the Place de la Concorde. 'It is a very exciting time to live in but people here are very quiet and refrain from celebrations till our main enemy goes under,'[22] he wrote to his mother, as the world awaited Germany's surrender.

It would take until the following year for the findings of the investigations he had revealed to Marsden to be published, carrying the deliberately unremarkable title 'Collision of Alpha Particles with Light Atoms'. However, the paper clearly proclaimed Rutherford's stunning conclusion that the atoms liberated by the collision between alpha particles and nitrogen 'are not nitrogen atoms but probably atoms of hydrogen ... if this be the case, we must conclude that the nitrogen atom is disintegrated'.[23]

Newspapers quickly distilled these findings to a blunt announcement. Tiny fragments chipped off by collision they might have been, but for the first time, a human had knowingly instigated the splitting of an atom. While this made for an arresting headline and a high point of scientific inquiry, the deeper implications of what the potential destruction of the atom might ultimately yield for the planet and its inhabitants were not immediately obvious – even to Rutherford.

He simply concluded that, such was the force at which those fractured hydrogen atoms were expelled upon the scintillation screen, the source of that energy must have been the nitrogen nucleus itself (complete with the binding energy that held particles together). Thus he offered the first experimental proof of one of physics' most familiar hypotheses: Einstein's theory of special relativity, which includes the famous equation '$E = mc^2$'.

More than a decade earlier, this theory had predicted the quotient at which mass might be transformed into energy, if such a means of disintegrating matter were somehow discovered. In Einstein's equation, 'c' represents the speed of light, a dauntingly huge number even before it is multiplied by itself. If the equation rang true, then the level of energy that might be liberated by splitting the atoms of heavy elements – those with the greatest atomic mass (Einstein's 'm') – was almost beyond comprehension. The heaviest naturally occurring element, as per Moseley's revised periodic table, was uranium, with an atomic number of ninety-two. The atomic mass – the collective sum of masses of the protons, neutrons and electrons contained in a single atom – of naturally occurring uranium's most common isotope is 238.

Until such time as Rutherford showed that the atom was not an indivisible entity, Einstein's premise had existed on paper alone. Now the urgent aim of experimental physicists worldwide would be to devise more and more powerful means of exploding matter's building blocks.

Rutherford would lead that charge, albeit within a different – if familiar – environment. At the height of his post-war fame, he was approached and appointed as Cavendish Professor of Physics at Cambridge University, taking over from his sixty-two-year-old mentor, J.J. Thomson.

To probe the inner-most secrets of the universe, Rutherford would build a team of the sharpest, most energetic minds that science had ever seen concentrated within a single laboratory.

6

A BENEVOLENT LORD

Cambridge, 1919 to 1927

Rutherford's return to Cambridge as Cavendish Professor in 1919 brought him a sense of muted triumph. While the laboratory had maintained its reputation as a centre of excellence in experimental physics under J.J. Thomson's continued stewardship, like most British and European institutions it bore significant scars from the Great War. Research had effectively ground to a halt as a bulk of the students and many staff rallied to the patriotic cause. A large experimental room had been reassigned to billet serving soldiers, and the technical workshops had turned out gauges for use in armaments, rather than vacuum tubes and precision instruments.

Rutherford had also wrestled with the notion of abandoning Manchester, where so many experimental triumphs had taken place, and where the Rutherford home had become a hub of social activity for physics department staff and students. Rutherford confided as much to his mother, after his appointment was confirmed in April 1919.

> It was a difficult question to decide whether to leave Manchester as they have been very good to me. But I felt it probably best for me to come here, for after all it is the chief physics chair in the country and has turned out most of the physics professors of the last 20 years.

It will of course be a wrench pulling up my roots again and starting afresh to make new friends, but fortunately I know a good few people there already and will not be a stranger in Trinity College. The latter will no doubt offer me a Fellowship which will give me the rights of the College to dine there when I please.[1]

The other factor that dulled Rutherford's excitement at inheriting the most prestigious post in global physics was the potential it carried to damage his relationship with Thomson, who was effectively being shunted aside. As he was considering the approach made to him by Cambridge, Rutherford wrote to his greatly admired former instructor: 'If I decided to stand for the post, I feel that no advantages of the post could possibly compensate for any disturbance to our long continued friendship or for any possible friction, whether open or latent, that might possibly arise …'[2]

Rutherford's angst was greatly smoothed by the reassuring response of Thomson: 'If you do, you will find that I shall leave you an absolutely free hand in the management of the Laboratory …'[3]

That was not how it panned out in practicality, however; Thomson would request that a number of his previous privileges be retained to allow him to continue research work in the manner he had enjoyed.

* * *

In preparation for their move from Manchester, Mary Rutherford and seventeen-year-old Eileen travelled to their new home town so Eileen could sit a senior school entrance exam. As they strolled across the grassed Backs, they ventured upon an abandoned house on Queen's Road that was being rapidly colonised by its voracious garden. Emboldened by the obvious neglect, the women broke in, and Mary felt an immediate affinity with the place.

When the family shifted to Cambridge months later and found it still unoccupied, Mary insisted that Ernest lease it from its owners, Caius College, and give it a major rebuild and refurbishment. The large garden, featuring a vast lawn that unfurled beneath sprawling trees and a mulberry bush along one of its distant perimeters, was a luxury virtually unknown in the heavily populated suburbs of Manchester. Mary wanted the grounds fastidiously maintained, which meant Ernest was compelled to take up a weekend hobby other than jaunts in the car.

As Rutherford reported to one of his former McGill colleagues, work was still going on early in 1920, almost a year after the family had settled in:

> We are in our home in Cambridge but still share it with the workmen. But we hope to have them out before long. It is New Year's Day and I have been exercising myself ... in sawing up a big tree in the garden so feel quite virtuous after a good day's work.
>
> I have thoroughly enjoyed my stay in Cambridge. The Lab has been crowded with both students and researchers but all has gone pleasantly and I hope efficiently. We are very short of room and must extend before long. I have been buying a good deal of apparatus to fill up the gaps and have been cleaning up the place generally.[4]

Requisitioning additional floor space and modest infrastructure for the Cavendish had been among Rutherford's first orders of business. He knew that realising the vision that guided his journey required an ongoing investment not only in technology, but also in personnel. He also reasoned that, with the war ended, 'the rising prestige of this country and the eclipse of Germany will lead to an increase in the number of research students from neutral and allied countries who wish to work in the Laboratory'.[5]

In addition to his requests for more places where physics could be taught and trialled, he called for the employment of extra lecturers, and for another chair of physics to be endowed. In the still-recovering social and economic climate, however, Cambridge University was unable to meet most of these demands, so Rutherford had no choice but to maximise the dividends of his own ingenuity and of the Cavendish's reputation. By so doing, he attracted a stellar team from throughout Britain and further afield.

Prominent among them were a charismatic Russian physicist (and another future Nobel laureate), Pyotr (Peter) Kapitza, and Englishman Patrick Blackett, the former commander of a naval destroyer whose ground-breaking sub-atomic photographs would feature in Rutherford's lectures. Under Rutherford, workspaces, ideas and results were freely shared at the Cavendish, and often robustly debated. This created a camaraderie rarely known amid the professional jealousies of other old-world institutions.

Wednesday afternoons were enshrined as the timeslot for the Cavendish Physical Society lecture, often delivered by a distinguished visitor to the laboratory. Rutherford would preside over these much-anticipated gatherings, and – having adopted British custom as firmly as she had embraced the class system – Lady Rutherford would also attend to serve tea. She had enjoyed resuming her Manchester custom of regular Sunday-afternoon teas in new surrounds at Newnham Cottage – and took even greater pleasure in putting students to work in her garden.

On Wednesday afternoons, the tea ritual was played out in the laboratory's poky Rayleigh Library, a space that smelled of stale biscuits and was not much suited to the sort of gathering that Mary liked to host. It was never made clear whether she was responsible for the book on etiquette that mysteriously appeared among the science texts on the library shelves.

By 1922, the irrepressible Kapitza had also founded his eponymous club, which brought the Cavendish's younger members together a few times each month, and often showcased a guest presenter who would bring members up to speed on the latest issues in physical science.

On one occasion, that speaker was Niels Bohr, whose rambling dissertation on German theorist Werner Heisenberg's complex 'uncertainty principle' prompted Rutherford to intervene and call a halt. The professor acerbically observed that his Danish friend's 'sense of time was as uncertain as the principle he was discussing'.[6]

Heisenberg, who later became a controversial figure due to his involvement in the Third Reich's atomic aspirations, himself addressed the Kapitza Club in July 1925. He spoke about the highly specialised topics of multiplets and Zeeman effects, and also answered questions on his recent research paper dealing with a new field that was changing the theoretical understanding of the universe – quantum mechanics.

* * *

Rutherford had quickly ushered in a new golden age of research at the Cavendish, despite being viewed as something of a patriarchal Victorian figure in both appearance and character by many post-war students. These were the days when university professors ruled their academic domain as a kind of personal fiefdom, yet Rutherford revealed himself as a benevolent lord.

Partly that was due to his disarmingly bluff persona. Those who had worked with Rutherford during his earlier incarnation at the Cavendish recognised the same loquacious character. As ever, his response to each exciting research outcome, or potential sighting thereof, was to burst into a preposterously loud, and even more tuneless, chorus of 'Onward, Christian Soldiers' as he marched off to his next assignment.

The descriptions of those most regularly in his presence paint a portrait uncharacteristic of a typical inter-war Cambridge don. One newspaper likened him to 'a hearty farmer who's enjoying his breakfast ... his healthy colour, blunt features, shrewd eyes, heavy limbs, and even his easy tweeds, with their baggy pockets, all seemed aggressively agricultural'.[7]

Charles (C.P.) Snow, who gained his physics doctorate at Cambridge under Rutherford in 1930 before pursuing a career in literature, would remember his professor as 'a big, rather clumsy man with a substantial bay window that started in the middle of the chest [as well as] large, staring blue eyes and a damp, pendulous lower lip'. Snow also evaluated Rutherford's New Zealand accent (via Montreal and Manchester) as 'bizarre ... it sounded like a mixture of West Country and Cockney'.[8]

For all his avowed commitment to updating Cambridge's equipment and outlook, Rutherford was reluctant to dispense with habits from the past. Financial conservatism was one of those traits – as was his slavish insistence that time be set aside each afternoon to undertake the counting of alpha particles. This practice had proven the cornerstone of his discoveries at Manchester, and remained a non-negotiable element of life under his rule at the Cavendish.

As Charles Ellis, who had foregone his chosen military career at war's end to pursue research work under Rutherford, later vividly recalled of that daily ritual:

> Counting the scintillations was difficult and tiring, and Rutherford usually had one or two of his research students to help him ...
>
> Sitting there drinking tea, in the dim light of a minute gas jet at the further end of the laboratory, we listened to Rutherford talking of all things under the sun. It was curiously intimate, but yet impersonal and all of it coloured by that characteristic of his of considering statements independently of the person who put them forward.[9]

* * *

In June 1920, barely a year after assuming the Cavendish professorship, Rutherford was awarded the Royal Society's annual Bakerian Medal, the second time he had been honoured with the venerable organisation's pre-eminent prize for physical sciences. He had previously earned the distinction in 1904 while at McGill, when his accompanying lecture centred on his discoveries relating to radioactivity. While the title of his second Bakerian Lecture – 'Nuclear Constitution of Atoms' – gave no hint as to its extraordinary prescience, it remains one of the oftest-quoted addresses in the award's almost 250-year history.

In his lecture, Rutherford explained in incremental detail how he had deployed alpha particles harvested from naturally occurring radioactive elements to bombard the core of a nitrogen atom until it forcibly released a hydrogen nucleus. It was, as had been revealed in his paper the previous year, the successful first step in the artificial transformation of matter.

However, he then went on – to stirrings among his audience, who sensed their orator was straying into the realms of scientific fiction – to make a series of predictions. These statements would doubtless have been enshrined as utterly fantastic had they not, within the span of fifteen years, been proven unerringly correct.

The first was to foresee a hitherto unknown form of hydrogen that carried double the atomic mass of the known element, and which would become integral to the processes of nuclear energy and weapons. (This material, when finally discovered in 1932, would gain the name 'deuterium'.)

The other was to float the notion of a neutral particle that would, if its existence could be proven, open the floodgates to the transmutation of matter, due to its capacity to circumvent existing electrical fields. The alpha bullets deployed by Rutherford and others were sufficient to penetrate the magnetic forces generated by nuclei of lighter elements. However, these fields were so strong that they protected hefty elements like uranium simply by deflecting any attack from charged particles.

> Such an atom would have very novel properties. Its external field would be practically zero, except very close to the nucleus, and in consequence it would be able to move freely through matter. Its presence would be difficult to detect by spectroscope, and it may be impossible to contain in a sealed vessel. On the other hand, it should enter readily the structure of atoms, and may either unite with the nucleus, or be disintegrated by its intense field, resulting in the escape of a charged hydrogen atom, or an electron or both.[10]

Rutherford referred to this figment of his unerring insight as a 'neutral doublet'. The world would come to know it, when it was eventually discovered at the Cavendish a dozen years thereafter, as the neutron.

In retrospect, Rutherford's conjecture proved a prediction of extraordinary acuity and courage. At the time, however, it was largely ignored, and soon forgotten beyond those in the immediate audience. To the staid scientific community, swayed only by peer-reviewed journals, it owed too much to speculation.

So, while the notion of the mysterious neutral particle remained in the back of Rutherford's fecund mind, the forefront continued to be occupied by the immediate concerns of the Cavendish Laboratory, and the responsibilities that came with being the world's foremost physicist.

As a result, the hunt for the 'neutral doublet' was instead taken up by Rutherford's most recent right-hand man, James Chadwick. A student at Manchester University, Chadwick – like so many others – had decided to abandon his mathematics course in favour of physics after he attended a lecture delivered by Rutherford in 1908.

He had been working in collaboration with Hans Geiger in Germany when war broke out, and was detained as he hurriedly tried to return to Britain. He then spent the entirety of the conflict at Ruhleben, a one-time horse-racing track in the Berlin borough of Spandau that became a British internment camp.

Although housed in stables, each converted to sleep six men on flimsy mattresses, Chadwick set up a crude laboratory in the freezing conditions and conducted experiments using German toothpaste, which contained mildly radioactive thorium powder. He also constructed a makeshift magnet from salvaged lengths of copper wire.

When released at war's end, Chadwick had returned – destitute, and in badly failing health – to recuperate in Manchester, where his father had worked as a railway porter. Learning of his plight, Rutherford had found him a part-time teaching position and then a studentship at the Cavendish. His kindness was repaid by the zeal with which the quiet, often withdrawn Chadwick took to his research task.

In the meantime, Rutherford himself continued to rewrite human understanding of the natural world. At a British Association for the Advancement of Science meeting held in Cardiff in August 1920, he suggested that the positively charged nucleus of the hydrogen atom should henceforth be called the 'proton'. The term was duly adopted.

During the following year, Rutherford enlisted his fame to help Vienna's Radium Institute, whose standing as the pre-eminent source of the highly prized radioactive material had suffered so severely as a result of the war and its political fall-out that the institute faced financial ruin. Having gratefully received a loan of radium from the institute while at Manchester, only for the element to become the annexed property of the British Government at the outbreak of war, Rutherford now set about raising money to reimburse the institute for its loss, even though his access to the raw material had been revoked.

In 1922, Rutherford was awarded the Copley Medal – the Royal Society's most prestigious honour, presented each year since 1731 for outstanding research achievement in any facet of science – for his examination of radioactivity and atomic structures. It placed him among exalted company that included Benjamin Franklin, Joseph Priestley, Charles Darwin, Louis Pasteur and – three years after Rutherford's elevation – Albert Einstein.

However, the event that brought as much pride and fulfilment as any peer recognition or laboratory revelation was the marriage of twenty-year-old Eileen to Ralph Fowler on 6 December 1921. The nuptials took place in the chapel of Rutherford's beloved Trinity College, where Fowler was a fellow. That, coupled with the knowledge that Fowler had attended Winchester College, famed for schooling Britain's social and intellectual elite since the fourteenth century, helped Rutherford to overlook the reality that his son-in-law was, by profession, a mathematician.

Rutherford's renown had become so great that in September 1923, the British Broadcasting Company decided it would air his presidential address to the British Association's annual congress, held in Liverpool. It was the first time a scientific lecture had been broadcast live into homes throughout Britain. It drew a generous response from informed listeners who found it stimulating to hear an expert in his field speak in depth on such a complex topic. A significant portion of the audience complained, however, that the subject matter was unintelligible and requested the exercise not be repeated.

Yet demand for Rutherford's appearances grew around the world, along with his reputation. In July 1924, he and Mary – accompanied by Eileen and Ralph Fowler – returned to Canada for a British Association meeting in Toronto. Afterwards, while the others took a horseback-riding trip through the Rocky Mountains, Rutherford repaired to a lakeside cottage in Quebec with his former McGill colleague (and subsequent biographer) Arthur Eve, where the men devoted their time to swimming, a few games of deck tennis and fly-fishing for speckled trout.

When an invitation arrived from the University of Melbourne's Overseas Lectures Committee suggesting a six-week lecture tour in 1925, comprising nineteen events in all six of Australia's capital cities, both Rutherford and his wife initially balked at the work and time commitment. A revised itinerary was then presented, comprising instead a three-and-a-half-week sojourn with speaking engagements in Adelaide, Melbourne, Sydney and Brisbane. The addition of a further two lectures in New Zealand sealed the deal.

'The primary intention is to see my people,' Rutherford candidly admitted to another of his former McGill University colleagues, Robert Boyle. 'But incidentally I shall give some lectures in the main cities of Australia and New Zealand.'[11]

In early September, hours after the couple disembarked in Adelaide and her husband attended a reception in his honour at the town hall, Lady Rutherford boarded the train for Melbourne, whence she would travel onward to Sydney then sail for New Zealand on the first available service. It was on the following day that Rutherford made his informal visit to the University of Adelaide's physics department that, unbeknown to him at the time, had such a consequential impact on Mark Oliphant, the young student who would soon revere Rutherford as both icon and idol.

When Rutherford set foot on home soil later that month, after similar receptions in Melbourne, Sydney and Brisbane, his arrival hailed the start of a triumphal return – one that the fêted favourite son somewhat sheepishly described as 'a semi-Royal tour'.[12] His itinerary had been meticulously planned by his former Manchester laboratory partner Ernest Marsden, now New Zealand's assistant director of education, and it included a series of civic receptions, free transport on any government-operated rail network and a private motor vehicle placed at the Rutherfords' disposal when staying at major centres.

Rutherford delivered lectures in Auckland and Wellington, in his former home town of Nelson – where he 'found the house where I was born at Brightwater had been removed and was now occupied by a chicken run'[13] – and in Christchurch. It was there, in the city where he and his wife had met, that the conquering hero was greeted by a stirring performance of a Maori haka presented by students from Canterbury College. Afterwards, the dance troupe illustrated the reverence in which Rutherford was held by manually dragging his vehicle, with Ernest and

Mary on board, to the city's municipal chambers for yet another formal function.

The reception afforded Lady Rutherford during other stopovers on their Antipodean tour was usually much less celebratory. These days she was often known to betray a brusque, overbearing manner; the colonial shyness she had initially betrayed after leaving Christchurch had given way to a burgeoning sense of entitlement as her husband's stature, and their resulting circumstances, flourished.

Having grown used to domestic staff as befitting Sir Ernest's elevated station, she bemused her husband's aged parents when she stayed with them at New Plymouth, where they had moved upon leaving Pungarehu. Each evening she would place her soiled shoes outside the visitors' bedroom door, and James Rutherford, now eighty-seven (and just years before his death in May 1928) would take it upon himself to quietly clean them. During that visit, she also insisted that Ernest's young nieces and nephews address her as 'Lady Rutherford', rather than simply 'Auntie Mary'.

* * *

Despite Rutherford's six-month absence, a steady flow of international students continued to find their way to the Cavendish, attracted by the laboratory's global repute. Among them was an unknown and uncertain American physics undergraduate, J. Robert Oppenheimer, who would earn immortality and infamy as mastermind of the atomic bomb.

Given that Oppenheimer's preference for theory contrasted with the Cavendish's charter as essentially an experimental laboratory, it was never likely to prove a comfortable fit. But so keen was Oppenheimer to embed himself within the intellectual pivot of world physics that he wrote directly to Rutherford seeking approval to work at Cambridge. Rutherford's lukewarm response, noting that entry to the Cavendish was restricted to those whose research capabilities were 'far enough along to work on a problem of their own',[14] was pejorative, though not completely dismissive.

With Rutherford on his southern hemisphere sojourn, Oppenheimer fell under the supervision of J.J. Thomson, who decreed that the man who would later oversee history's largest and most morally vexed scientific project must complete a basic course in laboratory training before being let loose on experimental apparatus. In return, Oppenheimer derided the lectures he attended at Cambridge as 'vile', and the practical instruction as 'a terrible bore'.[15]

Oppenheimer would ponder theory deep into the night before returning to the daily practical routine he found so distasteful, and became so run down and disillusioned that one day, mid-conversation with laboratory colleague Robert Ditchburn, he collapsed and was only spared injury by Ditchburn's reflexive save. 'It has been suggested to me that since I prevented him from cracking his skull on the floor I am responsible for the atomic bomb,' Ditchburn later mused.[16]

Months later, Oppenheimer was central to a more sinister episode when – struggling to cope in an environment where his theoretical brilliance was dulled by being but one of numerous science prodigies – he left an apple laced with traces of toxic chemicals on the desk of his tutor, Patrick Blackett. The poisoned fruit was left untouched, but Oppenheimer was placed on probation and instructed to undertake psychological counselling in London.[17]

History might also have taken a different turn had Rutherford been at the Cavendish to confront the erratic Oppenheimer. The professor with the occasionally incendiary temper bore no tolerance for petty academic squabbles, and would surely have taken a tougher stance.

Years earlier, upon joining the ceremonial line to enter Trinity College, where he routinely took Sunday-evening dinner, Rutherford overheard a curmudgeonly vice-master named Parry admonish a professor over some minor imperfection in the latter's academic gown.

'By thunder!' Rutherford railed at Parry once all had assumed their seats at the elevated top table, slamming his palm onto the polished woodwork. 'If you ever talk to me like that, I'll knock you base over apex down those stairs.'[18]

However, at the time of Oppenheimer's indiscretion, Rutherford and Mary were still enjoying their leisurely return from New Zealand, which included sightseeing stops in Ceylon (now Sri Lanka) and Egypt. When Rutherford did settle back into his office at the Cavendish, he was rarely in one place for very long. He threw himself immediately into further lecture tours across Britain and Europe, and between those engagements he reimmersed himself in the investigations that James Chadwick had continued making in his leader's absence.

'Chadwick and I are working at the scattering of alpha particles and hope to publish some interesting results before long,' Rutherford wrote to one of his former colleagues at the start of summer 1926. 'I want to know a little bit more about nuclei before I retire from actual work.'[19]

At that time, Rutherford remained oblivious to the impact that his Antipodean odyssey had exerted upon the young man who would soon become his trusted collaborator, avowed friend and anointed inheritor of his Cavendish legacy: Mark Oliphant.

7
'A RARE QUALITY OF MIND'

South Australia, 1919 to 1927

When Oliphant began as a full-time university student in 1919, the year of Rutherford's return to the Cavendish, he found himself among an influx of young men who had seen their lives disrupted, and often irrevocably changed, by four years of war. The only respect in which he stood out was that those who returned from active service had their fees covered by a grateful government, while boys like Oliphant deemed too young for the war effort had to pay their way.

Unlike so many of his later colleagues who had witnessed the First World War from hauntingly close proximity, Oliphant's memories of that great conflict would be tinged by an adolescent's sense of adventure. He had been preparing for his matriculation exams when the armistice was signed in 1918, and two of his uncles – Baron's brothers Walter (Instructional Staff Lieutenant) and Douglas (Private, AIF 10th Battalion) – had served. Both had returned safely home, which meant the war stories that Mark grew up hearing carried an uncommonly happy ending.

'It's funny, my generation felt deprived of an experience when the war finished,' he mused in his latter years. 'We'd had our uncles and brothers in it and we missed out. I felt deprived because the war ended before I was eighteen and involved. I can remember an uncle coming home after being shot in the leg and he was a great hero to us boys, listening to the stories he had to tell.'[1]

On starting his university career, Oliphant's initial vision was to utilise his science degree as a stepping stone to medicine, thereby satisfying the 'do-gooder' aspirations of his parents, who had accepted he would not be joining the Church. When he entered into further study, though, it became clear to him that the intensive theoretical focus of a medical degree held little appeal. As he knew from his schooldays, he preferred to be in the laboratory, 'fooling about' with hands-on experimental work.

'A medical course is almost entirely learning by rote, and there's nothing to think about in it,' Oliphant would later rationalise. 'You've just got to learn things, and I felt that this wasn't for me.'[2]

Oliphant's study choices were narrowing. He had been relieved of his position at the Adelaide Library because of his propensity to spend more time reading than re-shelving, and had paradoxically decided that medicine was not for him because it focused too heavily on reading. However, his skill with apparatus brought him to the attention of the University of Adelaide's physiology professor, Dr Thorburn Brailsford Robertson, who seconded the teenager's help for an experiment.

Robertson's aim was to glean whether animals derived any sustenance from nitrogen present in air, or if they could survive being fed and watered in an environment bereft of the gas. Having established that the mice would not be hurt, Oliphant employed his talent for designing and manufacturing equipment to ensure the generations of rodents were raised and monitored within a controlled atmosphere consisting only of oxygen and inert argon.

While the experiment itself yielded little of scientific value, its revelation was Oliphant's ingenuity in creating and maintaining a micro-climate in which the mice were kept at a constant temperature, with a strictly scrutinised combination of the appropriate gases. It also required regular removal of all waste products. Echoing Rutherford's formative years, the operating system for Oliphant's benchtop kit was powered by a small electric motor, driven in turn by a water wheel.

'The fun of building this equipment and finding ways of keeping the constituents constant, of removing the moisture that was produced, of introducing the food without introducing oxygen, of circulating the atmosphere and measuring it all the time by heat conduction to see what the relative proportions of oxygen and argon were, and automatically opening and closing valves, so fascinated me that I decided that physics was my milieu,' Oliphant would recall.[3]

By the end of 1920, and doubtless due to Oliphant's diligence, Professor Kerr Grant had offered the teenager a paid cadetship in the physics department. This required Oliphant to prepare and dismantle, clean and catalogue the apparatus used each day for first-year practical classes.

Kerr Grant had earlier gained fleeting insight into his new departmental cadet's character. Oliphant had once attended an Adelaide University Science Association meeting at which he had traded opposing views with the professor. The other students had looked on with a blend of disquiet and admiration as the pupil and his master firmly stood their ideological ground through a prolonged though entirely respectful debate. Later, Grant reputedly told a confidant, 'You know, despite his erroneous ideas and his immaturity, I believe Oliphant has a rare quality of mind. It could take him a long way.'[4]

While the laboratory assistant's role might outwardly have appeared to fall short of Mark Oliphant's ambitions, he was delighted to take up the position. Not only was the cadet's weekly salary of ten shillings equal to what he had been earning at the library, but the position came with a far greater benefit – namely, access to any course of his choice at the university, provided his duties in the physics laboratory were not compromised. The money he pocketed from his cadetship covered train fares to and from his family's new home at Cottonville (now Westbourne Park), with the surplus handed directly to his mother for board.

He was regularly co-opted into what he described as 'dog's body' jobs, yet he thrived on the practical and pragmatic stimulation of the laboratory, and his assuredness blossomed.

Now that experimental physics was his chosen discipline, it did not take Mark Oliphant long to embrace the field of research that would become his life's work. Still, the intricacies inherent in some of its complex calculations would regularly elude him due to his limitations in mathematics.

Initially that shortcoming was masked by his competence in experimental work. In second-year physics he topped his class, but as the subject matter became increasingly impenetrable in third year, he found himself struggling. Oliphant always acknowledged his scholastic limitations, claiming he chose to invest effort in the subjects that most interested him, while being content to secure the barest pass mark in all others. And there was no doubting that his keenest interest lay in practical work.

Nevertheless, he successfully completed his Bachelor of Science at the end of 1921. He was then offered the opportunity to undertake an honours degree the following year, which he accepted, and benefited hugely from being one of only two students in the physics department being personally supervised by Professor Grant. Upon graduating with first-class honours in 1922, he continued his research work into surface tension, which had become his then area of expertise.

However, as he expanded his experimental skills, his curiosity was excited by the pre-eminent physics conundrum of the early 1920s: the properties of radioactive matter. Oliphant accessed a jar of uranium salts from one of the chemistry department's supply cupboards, and joined the race to unlock the atom's secrets through his own amateur attempt at transmutation. Employing an induction coil that generated around 100,000 volts of electricity and a vacuum tube into which he placed his sample of the radioactive element, he bombarded the uranium with electrons for hour upon hour.

Fascinated when the university's chemistry professor, Edward Rennie, had explained that radium gradually decomposed to an isotope of lead, Oliphant aimed to initiate the same process in uranium by implanting additional electrons into its nuclear structure. That would – according to his narrow understanding of atomic theory – transform it into the isotope then known as 'uranium x' (later revealed to be uranium-234). However, the only noticeable change delivered by the lengthy experiment was to liquefy the overheated induction coil's insulating wax.

Rather than dampen his enthusiasm, this unsuccessful foray into experimental autonomy bolstered Oliphant's confidence. In the small university of barely 1000 students and a dozen professors, he continued to gain renown for the unusually quick and exceptionally skilful construction of apparatus he had exhibited since boyhood. He displayed a profound ability to craft experimental pieces, having learned the crucial art of glass-blowing from artisan Arthur Rogers who had originally travelled from England to work for William Henry Bragg, whose twenty-two-year tenure as Professor of Experimental Physics at Adelaide ended in 1908.

Rogers remained in Adelaide after Bragg's return to Britain, and maintained a lifelong correspondence with Oliphant who had emerged as a first-class glass-blower under such expert tuition. This gift also helped ensure that Oliphant's subsequent research career would be characterised by his reluctance to appreciate that others could not replicate the speed or intuition with which he built equipment.

And while 'the young man with mouse-coloured hair sticking up from a very large head'[5] – as one student contemporary described him – became wholeheartedly consumed by the world of academic possibilities, from the outset he was never motivated by personal reward or self-aggrandisement. His colleague and commuting companion Walter Schneider would reflect on this years later.

> I think Oliphant always had the spirit of the true scientist. I greatly doubt if he ever possessed any materialistic ambitions beyond a sufficiency of food, clothing, shelter and some relaxation. He was totally at ease with himself.
>
> His work was at the same time his reward, not something to be exchanged for dollars at piece rates, or by the hour. It was a free activity of his mind which he found fascinating, and sometimes dangerous, yet waiting to be done for its own sake – a drug for which most men have no taste. For him, and for a very few, it was … the meat and drink of life itself.[6]

* * *

Not that Mark Oliphant was oblivious to *all* activities beyond the laboratory. During his teenage years, he had met and become smitten with Rosa Wilbraham, a quiet, shy girl two years his junior. Neither could subsequently recall where they first met, but in later life, Mark would clearly remember the sight of Rosa's hypnotic dark eyes and flawless alabaster skin, framed by luxuriant dark hair that would cascade below her shoulders if not drawn into a tight bun.

They began to see each other regularly at church social gatherings, and before long Mark was foregoing weekend reading and backyard woodwork projects to catch the train then walk to Glenelg South, where Rosa – by then, tragically, an only child – lived with her parents.

Rosa's family quickly became fond of their daughter's new beau; later, Rosa gleefully claimed that her mother 'adored' him. Mark would arrive at the large Pier Street bungalow of Frederick Wilbraham – a printers' proof-reader who later worked as a correspondent for Adelaide's principal daily newspaper *The Advertiser* – bearing a tennis racquet and picnic lunch, before the teenagers set out towards a nearby park for the afternoon.

The conversations they shared between sets of tennis and bundles of cheese sandwiches would have been vastly different from those Mark engaged in during the week, given that Rosa professed no science training, and did not take up paid work upon finishing secondary school. Yet they spent all their available free hours in each other's company.

'We had the loveliest of times together,' Rosa would recall of their courtship, which included regular weekend outings such as formal dances and group picnics. 'Mark was the most beautiful dancer.'[7] On the spring day in October 1920 when Mark turned nineteen, they took the train and spent hours together in his beloved Adelaide Hills. Come Rosa's twentieth birthday two years later, he proposed as they sat together on the trunk of a fallen tree.

They were married at St Peter's Church in Glenelg on 23 May 1925. Rosa had celebrated her twenty-first birthday some months earlier; Mark was twenty-three. Using the smithing skills he had developed as a teenager, Mark crafted his wife's delicate wedding band in the university laboratory, from the small gold nugget that his father Baron had unearthed from the Coolgardie fields. In sixty-two ensuing years of marriage, that ring left Rosa's finger just once.

It was just three months later that Ernest Rutherford sailed into Adelaide and unknowingly recast the couple's future. Mark Oliphant's previously benign aspiration to pursue the life of a lecturer or experimenter was jettisoned virtually overnight, and his wife was swept along by his new-found ambition.

* * *

Yet the fire that Ernest Rutherford had ignited within Oliphant would ultimately smoulder and turn cold if he could not find a course of original study, and then a suitable means of getting to Cambridge. The solution to the first part of that problem became clearer at the start of the following year, when Professor Grant left Adelaide University to begin a twelve-month sabbatical at the laboratories of Thomas Edison's General Electric Company in Schenectady, New York.

In Grant's absence, Oliphant came under the guidance of the physics department's acting head, Dr Roy Burdon, who – at the professor's behest – had begun experimenting with the properties of surface tension, particularly in mercury. Burdon enlisted assistance from Oliphant, who had risen to the position of physics demonstrator and lecturer after completing his postgraduate research.

Even with his first-class honours degree, Oliphant remained unconvinced by his own academic acumen. 'I think I was rather lucky to get this distinction,' he conceded later in his life. 'I'm quite sure now that I was nothing like as good as the students that I've taught in the many years that I've spent in university teaching since. But still, I did get this great uplift that came from earning something for myself and went on to a little research work in the spare time that I had from my duties as an assistant in the laboratory.'[8]

As part of that research work under Burdon's guiding hand, Oliphant helped design and fashion apparatus for the mercury examinations, in which he would become closely involved. It also became his task to purify mercury samples, and he spent hours boiling and distilling the metal while its toxic vapour rolled down the stairs and seeped into the university's basement workroom.

The pair then went on to study the impact of other gases on the quicksilver's surface, and in 1927 they co-authored a paper entitled

'Surface Tension and the Spreading of Liquids'. It was the first time that twenty-six-year-old Oliphant's name had appeared in a scientific journal.

Oliphant would publicly air his gratitude to Burdon when his former physics instructor retired from teaching in 1959, noting the academic debt that he owed. He later added: 'I know in my heart that I have done nothing since as good as that first original work.'[9]

A follow-up paper appeared in *Nature* later that year, at which time Oliphant was encouraged by his senior colleagues to submit an application for the upcoming round of 1851 exhibition scholarships. Oliphant had now discovered a course of original study, and a scholarship would neatly solve the second dilemma he faced if he was to pursue a career at Cambridge: the physical and financial means of getting there.

Just as fortuitously as Rutherford, who had been among the first intake of 'outside' students in Cambridge's sheltered history thirty years earlier, Oliphant would benefit from the decision in 1928 to extend the 1851 scheme and deliver between four and six Australian scholarships, rather than a solitary annual award.

But while Oliphant's second published paper provided the foundation upon which his scholarship application was built, the 1851 Commission's criteria extended beyond an aspirant's ability to commit their research outcomes to writing. They were also required to show 'proved capacity for original work', and have their prowess endorsed by judges boasting sound science pedigrees.

Those testimonials came in glowing form from both Kerr Grant and Roy Burdon. The former wrote that throughout his time at Adelaide University, Oliphant had shown 'an unusually wide and thorough acquaintance with theoretical physics and an altogether exceptional skill in experimental work'.[10]

Burdon's assessment was even more effusive. 'Mr Oliphant's position and duties have given him practical opportunity for the publication of independent research work, but he has shown himself both skilful and versatile as an experimenter and is the most skilful manipulator of glass and quartz that I have known.'[11]

In July 1927, Mark Oliphant received notification that he had joined the ranks of the less than fifty successful past applicants from Australia –

just three of them from the University of Adelaide's physics department – to gain an 1851 Research Fellowship. After excitedly breaking the news to Rosa and his family, his first course of action was to send an urgent cable to Cambridge to introduce himself to Rutherford, and to request that he be admitted to conduct postgraduate research at the Cavendish.

It would be years before Oliphant realised that Rutherford – having been elected as a commissioner for the 1851 scholarships in 1921, and appointed to the governing board three years later – had known of his application's outcome before Adelaide University and its successful candidate were informed. Rutherford's response to Oliphant's telegram was a simple pledge. He would reserve a place at the Cavendish for the following term, just months away from starting. It was therefore only a matter of weeks between Mark Oliphant learning of his future and when he embarked upon it.

The £250 annual stipend offered under the 1851 scholarship program (around AU$20,000 today) was almost adequate for a single student travelling abroad, but, as Rutherford's rival James Maclaurin had realised thirty years earlier, it was too little to sustain a married couple. However, there was no question of setting out on such a significant life journey without Rosa, who quietly adhered to her mother's advice that 'where your husband goes, you go'.[12]

The financial impost that Rutherford had faced in 1895 had been eased slightly in Oliphant's day by a deal struck with Australasian shipping agencies, which covered the cost of passage for successful scholars. Therefore, the only pause in celebrations came when Oliphant wondered how he might find the funds for Rosa's fare. When that money was raised, through contributions from the couple's euphoric families and friends, passage was booked and their voyage viewed as the belated honeymoon they had not previously found time or means to enjoy.

On 10 August 1927, as Mark nursed his personal references from Grant and Burdon as well as the original manuscript of his *Nature* paper, which he planned to present to Rutherford at their inaugural meeting, the couple began their first journey beyond Australia's shores. It would be a seven-week sea voyage to Liverpool via Durban, Cape Town and Tenerife, aboard the steamer *Ascanius*: the same vessel that had delivered Ernest Rutherford into their lives two years earlier.

8

STRING AND SEALING WAX

Cambridge, 1927 to 1928

Today the Cavendish Laboratory museum is set among verdant fields and grazing horses in West Cambridge. Along its whitewashed internal walls, the evolution of the Cavendish graphically unfolds in a sequence of group portraits, stretching back to the century before last.

Every year, save for those when war demanded they be employed elsewhere, staff and research students who passed through the Cavendish's oak gates – or the sleek reinforced glass entry after the relocation in 1974 – have posed for posterity. A tale of the laboratory's hierarchy, wordlessly narrated.

In the first, stilted photograph from June 1897, the Cavendish's then director J.J. Thomson is the centrepiece. His arms are folded firmly across his chest, and unease at the photographer's staged arrangement apparent in his eyes. Around him, ten young men adopt similarly forced poses. At the frame's far right, sitting raffishly astride a wooden chair with its backrest to the camera, is twenty-five-year-old Ernest Rutherford.

Three decades on, the group has swelled more than fourfold, with Rutherford, as Cavendish Professor, occupying equal central billing alongside Thomson. The tall, studious figure of the recently arrived Mark Oliphant can be seen at the far end of the third tier.

Come the laboratory's 'golden year' of 1932, Oliphant has gravitated to the second row behind Thomson. Two years later, he has reached the prestigious front seats.

The final physics research students photograph to feature either Rutherford or Oliphant was taken in June 1937, with the pair prominently positioned front-row centre, separated by Nobel laureate Francis Aston. By then, Oliphant was effectively Rutherford's deputy, and their physical proximity in that poignant image reflected their essential closeness.

* * *

That day remained a decade away as Mark Oliphant hurried through Cambridge's narrow laneways on the first morning of Michaelmas term 1927. It was not only his attire – the sole tweed suit he possessed contrasting starkly with the billowing academic gowns of the cyclists around him – that told Mark Oliphant he had ventured far from Adelaide. As he turned into Free School Lane, the dirty flintstone wall above him offering no shelter from the softly drumming autumn rain, marked the rear border of Corpus Christi College – established around 500 years before South Australia was colonised.

Such was his haste to take refuge within the Cavendish's covered entryway that he had no time to absorb the symbolism of that moment. It was the culmination of a journey two years in hopeful planning, and almost two months in physical execution.

Had there been opportunity to pause and peer into the low, grey sky, he might have noticed the figure of the seventh Duke of Devonshire, William Cavendish, set in stone above the entrance. The Latin lessons Oliphant had taken to satisfy his father would have helped him decipher the accompanying inscription: *'Magna opera domini exquisita in omnes voluntates ejus'*, a verse lifted from the Latin Bible that translates as 'The works of the Lord are great; they are studied by all who delight in them.' It was as near as the world's best-known science research facility came to street signage.

As Oliphant entered the passageway and shook the droplets from his broad shoulders and thick, wavy hair, he received a gruff direction towards three stone steps that led to another set of forbidding doors, whence a wooden staircase would take him to the Cavendish's heart.

What the front-gate porter did not warn him about was the tangle of bicycles that lay strewn immediately beyond the doors, parked within the secure confines of the Cavendish's ground-floor corridors as an anti-theft measure.

'Everybody had parked their bicycle there so they wouldn't be stolen,' Oliphant later recalled, noting that he was forever being relieved of his wheels when he visited places other than the Cavendish. 'A bicycle left in the street was free-for-all. I was continually going to the police station to pick out my bicycle from the hordes of bicycles that had been picked up by the police. Somebody would "borrow" it and go from A to B, and then just abandon the bike.'[1]

Now Oliphant picked a path through the mess of metal before eyeing a wooden staircase that groaned and squeaked under approaching footfalls. The Canadian research student who loomed into view was delighted to find another Commonwealth colleague, and guided Oliphant up the narrow steps and into an undulating corridor overlooked by sepia portraits of the previous Cavendish professors – James Clerk Maxwell, Lord Rayleigh and J.J. Thomson.

As Oliphant followed the young man, who had introduced himself simply as Henderson, he noticed the thick film of dirt that coated the building's tall, Gothic windows, minimising the already muted light that penetrated the dank hallways. 'At first I was a bit upset by the fact that everything was so filthy,' Oliphant mused much later. 'I discovered in the end that the windows ... and the floors were washed once a year.'[2]

A climb of several more stairs brought them to the Rayleigh Wing. Upon reaching a bare pine door at the end of a short passage, Oliphant was advised to wait until the famous physicist on the other side was ready to receive visitors. The churn in his stomach was not dissimilar to the mild nausea of seasickness that he had recently endured during the six-week voyage from Adelaide, but which had subsided after he and Rosa had changed to the 'boat train' that connected the Liverpool docks with London. The couple had stayed there a few days with relatives of his wife's.

Now, as he loitered distractedly in the corridor awaiting his summons from Rutherford, Oliphant found himself in the company of two other 1851 scholarship recipients. One was Cecil Eddy, recently arrived from Melbourne; the other, from across the Irish Sea in Dublin, was Ernest Walton, whose name would become forever associated with one of the laboratory's most famous finds. Walton was in even greater turmoil than the two Australians, having confused his paperwork and arrived too late for his proposed meeting the previous morning. All three nervously anticipated their first audience with Rutherford – memories of which would remain forever vivid for Oliphant.

> When my turn came, I entered a small office littered with books and papers, the desk cluttered in a manner which I had been taught at school indicated an untidy and inefficient mind. It was raining, and drops of water ran reluctantly down the grime-covered glass of the uncurtained window. I was received genially by a large, rather florid man, with thinning fair hair and a large moustache, who reminded me forcibly of the keeper of the general store and post office in a little village behind Adelaide where I had spent part of my childhood.
>
> Rutherford made me feel welcome and at ease at once. He spluttered a little as he talked, from time to time holding a match to a pipe which produced smoke and ash like a volcano. Later on, I found that he reduced his tobacco to tinder dryness on a newspaper spread out before the fire at home, or on a radiator in the laboratory.[3]

That Oliphant instantly saw in Rutherford a likeness to Mr Cooper from Mylor's grocery-cum-everything store offered greater reassurance than did Rutherford's handshake. Oliphant found it, in contrast to the professor's boisterous tone, 'very brief, limp and boneless ... He gave the impression that he was shy of physical contact with another person'.[4]

No such reservation emerged from the discussion that followed. After preliminary niceties, Oliphant produced from his jacket's inside pocket the manuscript of his published work on the absorption of gases upon freshly formed mercury surfaces. Having studied the results that Oliphant had gained with Burdon in Adelaide, Rutherford took a deep drag on his pipe, blew more sulfurous smoke into the already foul air and announced: 'Well, you can start work at once,' before adding: 'I suppose you would like to know what you'll be doing?'

Oliphant then detailed the experiments he had in mind, which involved the effects produced when positively charged ions struck the surface of various metals, an extension of the work he had earlier undertaken using mercury. The results he hoped to gain, Rutherford acknowledged, could help inform experiments being undertaken elsewhere in the Cavendish to examine the discharge of electricity through gases. Oliphant then reached back into his pocket to remove a further sheaf of folded papers, upon which he had drawn diagrams of the apparatus he would need to undertake that work – most of which he could construct himself, of course.

Rutherford beamed through the fog from the other side of his disorderly desk. 'Well you can go ahead alone,' he roared, prising himself from of his chair. 'By the way, go 'round the lab and talk to the boys. You might start with Aston and Chadwick, and if J.J. is in, he would be interested in your experiment. You'll find him working in the garage, or in nearby rooms.'[5]

Oliphant hesitated in the doorway, as he prepared to re-enter the network of gloomy corridors. The prospect that he might come face to face with Sir J.J. Thomson, Nobel Prize winner and discoverer of the electron, or Francis Aston, who had become a Nobel laureate five years earlier for discoveries using his elegantly brilliant mass spectrometer, was altogether too daunting.

As he turned to clarify with Rutherford where he might find this 'garage', which seemed a repository for scientific royalty, the professor dismissed him with: 'Now don't be diffident. Tell them I sent you.'[6]

His brain still whirring, Oliphant almost collided with two young men who barrelled out of a room on the opposite side of the corridor. One of them flashed a broad smile, thrust out his hand and announced, 'I'm Blackett. Who are you?' Oliphant had already gleaned that the Cavendish's members addressed one another by surname alone. Blackett then offered to take the new boy to where Rutherford's world-renowned older 'boys' continued to reshape comprehension of the natural world.

'This was my introduction to Rutherford and the Cavendish Laboratory,' Oliphant would write years later. 'Rutherford's warm welcome and interest, combined with my conviction that I could do better glass-blowing than J.J.'s assistants were able to accomplish, gave me courage. The Cavendish and Cambridge were already becoming part of me.'[7]

* * *

While it's scarcely discernible in the current Cavendish Museum's monochrome images of mainly fresh-faced men, with a scattering of women whose ambitions Rutherford actively encouraged, Oliphant was a few years older than the typical novice research student. He was about to turn twenty-seven, and fully aware that, by the same age, Rutherford had written his name permanently into scientific lore by identifying and naming two distinct forms of radiation – the first in a litany of quite astonishing breakthroughs during the decades that followed.

Within the pool of brilliant young minds that had subsequently followed Rutherford to Cambridge, competition to unearth the next great find was understandably fierce. Which meant that Oliphant was not about to mark time before getting down to business.

In his favour were the ease and intuition with which he could fashion and utilise equipment. These skills, coupled with his willingness to work, meant that he immediately gained Rutherford's approval. 'I want you to go with Oliphant,' Rutherford once counselled Philip Moon, who would spend much of the next decade or more working alongside the Australian. 'He's a very fast worker.'[8]

As a consequence of his obvious abilities, Oliphant was not required to complete an apprenticeship in the 'nursery' loft, set beneath the gabled roof among the rarely visited individual offices of various senior staff, including the irascible Charles (C.T.R.) Wilson. This was where his fellow 1851 scholarship holder Walton found himself immediately stationed, and where Oppenheimer had endured six months of anguish during his time at the Cavendish two years earlier.

Instead of the laboratory finishing school, Oliphant was allocated a workspace in the Rayleigh Wing, under the nose of Rutherford and cheek by jowl with half a dozen other bright young researchers. It was at a benchtop in the corner of this large, upstairs room, which hummed to the tune of gas jets and electrical equipment, that Oliphant began his Cavendish journey.

Initially on his own, and then in collaboration with his research student Moon, he pursued the study of positive ions as he had outlined to Rutherford. He devised new ways beams of those ions extracted from various elements could be accelerated by applying modest electrical charges of up to a few hundred volts. It was work that, over the next two years, would form the basis of his PhD thesis.

Settling comfortably into his preferred role as an experimenter and master equipment manufacturer, he quickly became part of the Cavendish fraternity, which was as eclectic as it was tight.

The most influential characters in his Cavendish life, aside from Rutherford, were James Chadwick and John Cockcroft. At the time of Oliphant's arrival in Cambridge, Chadwick was Rutherford's closest experimental ally. The pair were engaged in disintegrating the nuclei of lighter elements by bombarding them with Rutherford's favoured artillery, the alpha particles given off through natural radioactivity. These particles were obtained from a precious radium source that Rutherford would invariably find the means to procure.

Still visited by regular ill-health as a result of his German imprisonment, Chadwick was an intensely shy, often abrupt figure, whom Oliphant initially found somewhat intimidating. As Rutherford's deputy and trusted confidant, Chadwick kept an eagle eye on all that unfolded in the cramped, often untidy laboratory, and frequently visited Oliphant at his workbench.

On one occasion, peering over the top of his round glasses, Chadwick inquired if there was anything Oliphant might need to aid his experimental progress. Seizing the chance to request a motor-driven Hivac vacuum pump to replace the comparatively inefficient hand-driven models, Oliphant added to bolster his case: 'that would make life much easier'.[9]

'Well there's none to spare,' Chadwick snapped back, then turned on his heel, leaving Oliphant shocked and fearful that he had caused mortal offence. When the young student returned from an extended lunch break, spent stewing over how he might mend the rift he had clearly caused, a Hivac pump awaited on his workbench, with neither explanation nor recrimination.

Oliphant's relationship with Cockcroft was far more straightforward, and immediate. Cockcroft had been a late convert to physics, having completed a degree in engineering and then studied mathematics at Manchester University before the outbreak of war. Like so many who formed the Cavendish team, Cockcroft became attracted to physics on the strength of Rutherford's lectures. He too had arrived at Cambridge – after a cadetship with electrical engineering heavyweight Metropolitan-Vickers – through an 1851 scholarship. He was also a newlywed, having married Elizabeth barely a month before Mark and Rosa Oliphant settled in England. Thus began a lifelong friendship between the two couples, with the Cockcrofts lending invaluable support and pastoral care to the young visitors as they adapted to life in Cambridge.

Undoubtedly the most recognisable figure at the Cavendish was Thomson, now seventy, who would regularly appear to tend to his ongoing experimental work. Just as often, however, he could be found strolling slowly and splayfooted through Cambridge's narrow streets, with his walking cane clasped in two hands behind his back and posing a hazard to passing pedestrians when he leaned forward to peer in shop windows. The grand old man would still occasionally deliver lectures, during which he was known to pause mid-sentence and flash an enigmatic smile, later revealed to be an urgent readjustment of his dentures.

It was Thomson who had come to institute the 6pm closure of the Cavendish on weekdays. When the clock struck six, everyone was compelled to vacate as the porter, oversized keys clanking in his hand, theatrically pulled shut and locked the heavy oak gates. Working on weekends was also forbidden. Despite his early career habit of toiling late into the night, Rutherford had maintained the tradition. He would announce to all those still bent over their benches as the mandated closing time neared: 'If it's not concluded by 6pm, you're better off going home to think about it.'

During working hours, however, movement was more fluid. The laboratory thrived on its open-door policy, whereby staff and students were encouraged to wander freely into research spaces and common rooms to share ideas and learn first-hand. It fostered a palpable sense of teamwork, and helped fuse together the formidable intellectual force that Rutherford had assembled. This, in turn, would deliver a decade of most remarkable scientific output.

Oliphant's other new colleagues included the testy Aston, who worked alongside Thomson in the basement 'garage' and whose lectures were rated among the most interminable, being lifted directly from a textbook. But he would prove an unstinting source of wisdom to the eager young researcher from Adelaide.

Arthur Eddington's presentations on relativity were much more accessible, and it was his championing of Einstein's formulae in the previous decade that had helped embed physics, and its explanations of matter and the universe, into the wider public consciousness.

Patrick Blackett's images of disintegrating nuclei brought to life the stories of the sub-atomic world that Rutherford presented across the globe, using photography made possible by the visionary cloud chamber, designed and then enhanced by another Cavendish character, C.T.R. Wilson.

Like Rutherford, and then Oliphant, Wilson had been drawn to science through his close connection to nature. An inveterate hiker, he was working on weather observations on the highland peaks of his native Scotland when he noted the halo effect of sunlight upon the mist of low cloud that settled around him. From there, the idea of an instrument that might recreate this phenomenon in a laboratory setting took shape.

Wilson found that if he suddenly expanded moist air within a sealed vessel, the moisture would condense into a mist of tiny drops, and this allowed him to recreate the glories and corona that are visible around the sun and moon. It was the first time a cloud had been artificially created in a laboratory, and in turn the cloud chamber would become an invaluable tool for showing the passage of highly energised sub-atomic particles.

Rutherford liked to joke that, as he departed the Cavendish for Montreal in 1898, he went to Wilson's workroom to bid him farewell and found the taciturn Scotsman studiously grinding the rounded end of a piece of glass tube into the shoulder of a second length. When he next returned to the laboratory years later for a visit, Rutherford vowed he went to the same room, which revealed Wilson at the same bench, grinding together the same pieces of glass valve. Wilson would never actively refute the tale.

Charles Galton Darwin, grandson of the revolutionary evolutionist whose book had led Oliphant to a life of science, was among the institution's other imposing names, as was Ralph Fowler, the Cavendish's resident mathematician, who came to occupy the office alongside that of Rutherford – his father-in-law – on the second floor.

* * *

The other influential personality, if not a direct presence, on the Cavendish team was Mary Rutherford. Her most famous contribution to the life of the laboratory was the tea gatherings she would routinely host at Newnham Cottage on Sunday afternoons. These events were notionally social occasions for resident and visiting members of the Cavendish, but in truth they were planned and executed as Lady Rutherford performance pieces.

Rather than casual get-togethers at which to mingle away from the workplace, the regular occasions proved more formal than any weekday routine – from the handwritten invitations stipulating an iron-clad 4.30pm arrival, to Mary's unilateral announcement of the conclusion, after which she expected a note of thanks to be sent by all attendees. As Oliphant later remembered, the dénouement of Sunday afternoons at Newnham would follow an inevitable course.

> An hour, or a little more later, if it was fine outside, we were asked by Lady Rutherford whether we would like to see the garden, and after a stroll we were led firmly to the door in the outer wall [which opened directly onto Queen's Road], where we shook hands and

departed. If the weather was bad or the light had gone, the end of the occasion was signalled by Lady Rutherford rising and going around the circle bidding us goodbye. There was no opportunity to outstay our welcome.[10]

The running sheet for these gatherings was also rigidly prescriptive, and scheduled to allow Ernest his regular Sunday-morning golf round at the nearby Gogmagog Hills course. Sunday-best attire was compulsory for afternoon tea, and guests were greeted at the cottage's front door by a maid in full Victorian chamber outfit, complete with ankle-length apron and white, lace-fringed cap. They were led past Rutherford's study – a clutter of books and papers that mirrored his Cavendish office, and which Mary would ruthlessly tidy when it became too great an eyesore – and into a large drawing room that overlooked the slavishly maintained garden. The room was dominated by a concert grand piano, which reflected Mary's musical preferences, given that her husband's tastes tended more to military-style marching bands.

Each visitor was then welcomed by the hosts, and directed to their appropriate place among the arc of prearranged seats. When all were in place, Mary would take up her post on a low chair before a diminutive table and pour the tea from a silver pot. It was a keepsake of which she was demonstrably proud.

Ernest would sit on the opposite side of the circle and lead the conversation, most often centred on world events, with a decided emphasis on news from the Commonwealth, gleaned from reports in *The Times*. Shop talk from the Cavendish was not permitted unless a quiet confine could be found beyond Mary's earshot. Rather, the guests assembled in the drawing room, balancing fine china cups and saucers upon their laps while juggling delicate triangular sandwiches, listened closely as Rutherford marshalled the chat, which was usually interrupted only by his wife's regular scolding.

'Ern, you're dribbling,' she would shrill, or she would rebuff his request for a second cup by noting that Mr Oliphant, or some other guest, 'hasn't finished his first yet – so you will wait'. The Rutherfords maintained separate bedrooms at Newnham Cottage, and there was never any physical affection shown between them, but those who saw them together often claimed that they were clearly devoted to one another.

'I don't know that she was a great conversationalist,' Elizabeth Cockcroft would recall of Mary Rutherford. 'People found her difficult to get on with, really.'[11]

There was a distinct duality to Mary Rutherford's manner. It appeared she had developed a strong, combative personality after being brought up by a strict, strident mother and a father who was absent, first through his drinking, then through his premature death. When in the company of Britons, Lady Mary would adopt the countenance of a society matron in keeping with the status her husband's reputation afforded. However, her demeanour was known to soften markedly when she was providing support and assistance to others – particularly those who were new to England and unfamiliar with its entrenched, class-based customs, like Mark and Rosa Oliphant.

Upon learning, in 1927, that the latest colonial addition to her husband's laboratory team had brought with him a shy, equally unworldly wife, Mary Rutherford insisted that the Oliphants become fixtures at Sunday-afternoon tea. With access to her own motor car, she would also routinely offer transport to Rosa Oliphant. Like Elizabeth Cockcroft, who formed a close bond with Rosa, Mary Rutherford took a shine to the young, financially struggling couple from distant South Australia, whose story she herself had lived.

* * *

The language, the signage, the cultural and culinary idiosyncrasies of England had been reassuringly recognisable to the Oliphants from the moment they alighted in Liverpool. Cambridge, however, with its medieval rituals and stifling formality, was like nothing they had seen, or prepared for. For that reason, as well as the financial constraints imposed by Mark's scholarship income, the early years at the Cavendish brought numbing hardship for the young couple.

Among Cambridge's protocols was the requirement that all students join a college, provided their chosen institution accepted their application.

Oliphant opted for Trinity College, founded by Henry VIII in 1546, which listed Newton, Maxwell and Rayleigh among its notable alumni. Rutherford had been a member during his first stint at Cambridge, as was J.J. Thomson, who in 1927 was Trinity's master.

However, college accommodation was restricted to bachelors, and it took some time for the Oliphants to secure modest lodgings suitably close to the Cavendish. In the interim, they made do with temporary accommodation.

Having spent their early married life in a self-contained wing of Rosa's family's large bungalow in the beachside Adelaide suburb of Glenelg, the shock of their new English digs never fully subsided. Their first flat, upstairs on Bateman Street, comprised a bedroom, a small living space and a kitchen that doubled as bathroom. Their kitchen table became a length of board balanced across the bathtub's edges; the couple had rejected the option of a 'hip bath' that could be dragged in front of the fireplace for the ritual Saturday night wash.

If the conditions were trying, the poverty was worse. Mark's scholarship granted him £250 per year (around £15,000 today), and when £2 went to weekly rent alongside the annual cost of college fees (£80) and textbooks (£20) it meant less than £50 each week in today's currency for living costs. Rosa's attempts to find work were stymied by her lack of qualifications in typing and stenography, the most marketable skills for women in inter-war Cambridge. So instead she set her mind to balancing a fragile household budget. 'Now dear, it's your job to make it go round,' Mark would instruct his wife as he handed her their meagre living allowance.[12]

Rosa would stoke the coal fire into life each day, shortly before Mark made the 1.5-kilometre walk from the Cavendish for lunch. Her goal was to create the appearance of a warm, welcoming home for her husband, only to extinguish the fire the moment he returned to work. Coal was expensive, so Rosa spent the hours either side of the midday meal huddled in an overcoat as she went about her domestic chores. On Sundays, that list included dry cleaning Mark's only suit.

So bleak was the financial situation in which the couple found themselves that, with their first English winter looming, Oliphant saw no option but to apply for the 1851 Commission's hardship provision: a supplemental payment of £30, to cover his tuition and other associated costs.

The memories of these painful early years in Cambridge would forever remain with Rosa, her reticence to spend on non-essential extravagance a characteristic she maintained even as her husband's career bloomed.

For Mark, the experience was vastly different. He came to cherish the years he spent at the Cavendish, and not only because of the camaraderie within the laboratory where, despite the evening and weekend curfews, Rosa felt he invested too much time. He also enjoyed the backslapping of Trinity College dinners held in the Great Hall, took up squash – which he played sporadically and unspectacularly in preference to offers from Cockcroft and Blackett to join the laboratory's hockey team – and would occasionally splurge on tickets to amateur theatre performances and music recitals. Mark and Rosa's classical tastes were well catered for in a community so strongly steeped in British high culture.

These rare indulgences, however, were contingent on more than just household expenses and college fees. Rutherford's steel grip on the Cavendish's expenditure meant that students were, at times, expected to pay for essential items themselves – forcing Oliphant to dip still further into Rosa's overstretched household purse.

* * *

If Oliphant had imagined, through the years in which his hopes of working alongside Rutherford had fermented, that the Cavendish was some gilded scientific nirvana, he had been disavowed of that idea early. Suggestions that it had evolved little from Thomson's days were, if anything, generous: others who preceded Oliphant into its hallowed workrooms felt it remained a timeless relic of Maxwell's original vision.

Part of its problem was cast in the sculpted flourishes of its minimalist street frontage. The entrance off Free School Lane might have served as a suitably grand statement when designed fifty years earlier, but now stood as an intractable impediment to the installation of large, modern machinery. The electricity supply from an equally antiquated power station on the banks of the River Cam was fluky, and so noisy motor generators had been set up outside the porter's lodge to feed the laboratory, and putrefy the air. This, in turn, helped ensure that much of the machinery at the cutting-edge research facility continued to be hand-operated. It wasn't only through caricature that the Cavendish was regularly spoken of as a bastion of 'string and sealing wax' experimentation.

Even as he familiarised himself with the workings of the 'garage' and then the upstairs 'nursery' – where novice research students were let loose on obsolete or even broken apparatus to guarantee any missteps did not incur costly damage – Oliphant was struck by similarities to his uncle's backyard repair shop, and the adolescent experimental retreat he had set up beneath the family home at Mitcham.

Among the purpose-blown glass flasks and precision-made brass calibrators were biscuit tins, some of which housed still smaller cocoa tins to keep vital parts and reagents separated. Rather than using commercial alternatives, the vacuum grease employed to fix stopcocks and ground joints was hand-made through a laborious process that required days of patiently cooking crepe rubber retrieved from inside the hard casings of golf balls. Constant stirring over low heat was required to achieve the desired consistency.

Vacuum tubes, so essential in the study of ionised gases, underwent initial evacuation through use of a hand-cranked pump, with the residual gas then absorbed by charcoal cooled through the application of liquid air. This was not the 'activated' charcoal that could be purchased, but the Cavendish version, which was made by carbonising coconut shells. And while the trademark crimson Bank of England sealing wax was prized for securing airtight fittings before the invention of plasticine revolutionised the practice, Cavendish members often preferred their own recipes, as prepared by J.J. Thomson's laboratory technician. The hard and soft versions of the sealant were both made from beeswax and resin among other ingredients, though the identities and proportions of each remained a close-guarded secret.

Overseeing this all-pervasive thrift was the Cavendish's chief laboratory assistant Fred Lincoln, an archetypal Dickensian caretaker with a collection of keys for every known cupboard and 'a moustache with wax-twisted ends which made him resemble a recruiting sergeant from the Edwardian era'.[13] Such was Lincoln's devotion to his task that he would stand at almost full salute when he encountered Rutherford on the workshop floor. He had received his start under Thomson, and had earned his stripes guarding the materials of everyday science with a zealotry usually reserved for micrograms of radium. He measured lengths of electrical cable and rubber tubing down to the prescribed inch, and hand-rationed issues of screws.

'He was shrewd, and had a good knowledge of human beings,' Oliphant later recalled. 'On one occasion, working after hours, we needed some tungsten wire. The relevant cupboard was locked, but we managed to pull it from the wall, take off the back, abstract the wire and replace the cupboard. We thought this had gone undetected, but when some time later we tried to repeat the performance, we found the cupboard screwed to the wall from the inside.'[14]

However, the parsimony that pervaded the entire laboratory stemmed from the man who loomed larger than life over every aspect of the Cavendish and its personnel: Ernest Rutherford. Raised in colonial poverty, schooled amid rudimentary facilities and with his most recent forays into academic administration tinged by the privations of war and then the gathering clouds of global depression, Rutherford preached frugality and self-sufficiency.

Having flourished as a researcher under Thomson in conditions even more austere than those he came to preside over, Rutherford felt that noble poverty produced more worthwhile results than pampered luxury. That view extended to seeking additional funds that might supplement the Cavendish's always inadequate budget. He was stubbornly unwilling to beg for assistance in any form, other than for his coveted supplies of radioactive source materials.

His gruff reminders as to the evils of excess had become legend by the time Oliphant arrived. 'We don't have much money, therefore we must think' was a refrain he trotted out with similar fondness to his grating renditions of 'Onward, Christian Soldiers'. While a research student at the Cavendish, Philip Moon found himself in need of liquid air as a coolant for work he was pursuing. On learning of another request lodged for that resource, Rutherford pronounced in his rumbling baritone: 'the time has come, Mr Moon, to decide whether the results you are getting are worth the liquid air you are using'.[15]

Furthermore, Rutherford maintained that his thirty or so research students would develop stronger investigative protocols and deeper experimental insights if they designed and built their own counters, amplifiers and power supplies, rather than simply plucking them from a manufacturer's shelf.

Baron Bowden, who completed his Cambridge doctorate in nuclear physics at the height of the laboratory's golden era in the 1930s, regarded researchers at the Cavendish as 'the most impecunious I've ever known'. Chadwick, Rutherford's trusted and loyal subordinate, once confided to a colleague: 'You would be surprised to know what this laboratory has been running on for the past few years, less than some men spend on tobacco.'[16]

Oliphant's examination of the properties of positive ions and the separation of isotopes through electromagnetism soon demanded a metal mercury diffusion pump superior in performance to the Hivac he had sought from Chadwick. His only options were to find the necessary £24 – almost ten per cent of his annual income – or proceed without. The pump was duly purchased. On other occasions, he was forced to buy a glass-blowing torch, secure supplies of rare gases, and pay for reprints of papers he had no choice but to type himself.

He chose to spend the money that he and Rosa could scarcely afford because science was everything to him. And at the core of that devotion lay the deeply personal nexus he had formed with his reason for being there: Ernest Rutherford.

9

A MEETING OF MINDS

North Wales and Cambridge, 1928 to 1932

The vista from the bank of windows set into the thick stone walls of the seventeenth-century Welsh longhouse, perched part-way up the eastern rise of the Nant Gwynant valley, seemed to stretch, unhindered, almost to New Zealand. Certainly, the spectral outline of Mount Snowdon, which loomed, treeless and mist-enshrouded, above the waterway that burbled across the valley floor, among emerald-green fields blotted only by disused slate quarries, was reminiscent of the volcanic hillscapes and high-country plains among which Rutherford had spent so many years, on the verdant South Island.

He had taken a lease on the whitewashed farmhouse named 'Celyn' (pronounced 'kaylin') soon after he received his Nobel Prize purse in 1908. The homestead – essentially incorporating the footprint of three stone cottages joined under a single roof to form one dwelling that historically housed both humans and livestock – was more than 200 kilometres from Manchester, where he had lived at the time. And the Wolseley roadster also financed by the award had allowed the Rutherfords to retreat to North Wales whenever Ernest's schedule allowed – at a minimum, for part of every university vacation period.

As Chadwick noted of his close collaborator's enduring love of nature:

> Rutherford was never so fully relaxed as when he was in the countryside. He was still a countryman as he had been in his youth … [and] he maintained that an active family should, every few years, leave the town and live in the country. His point was partly that one should get away from the tensions of urban life and so recharge one's batteries but chiefly that one needed to regain contact with nature and the country life on which man ultimately depends.[1]

'To one who knew the completeness of his immersion in his work while in the laboratory,' Oliphant would remember of Rutherford, 'his ability to put it aside and relax utterly while on holiday seemed remarkable. Moreover, he urged his colleagues and students to do likewise, often telling one who was obviously tired to go off for a complete break.'[2]

Motoring had increasingly become another of Rutherford's extracurricular joys. When he first took the Wolseley on the roads around Cheshire, near Manchester, he had delighted in pushing the vehicle to around twenty-five miles per hour (forty kilometres per hour). On his move to Cambridge, from where the journey to Celyn was closer to 350 kilometres, that top speed was regularly doubled. Those longer journeys required overnight stops with friends along the route, when Mary Rutherford would bluntly inform their hosts that she and Ernest would require sandwiches for their onward travels next morning. And upon receiving the order, would often peer into the contents and tersely complain, along the lines of 'sardines – don't like sardines – get me something else'.[3]

By the time the Oliphants arrived at Cambridge in 1927, advances in both highway infrastructure and automotive technology meant that the distance could be covered in a single, albeit long and not altogether comfortable, day. Oliphant later noted that there was little of aesthetic interest in the countryside once past Shrewsbury, heading west into Wales.

> It was a long and arduous drive. The car was what is known to Americans as a convertible, with hood which could be folded back behind the seats in good weather. There was no heater or windscreen

wiper, and when in winter it was at times necessary to drive through fog or a snowstorm, with the windscreen open, it was bitterly cold.

Lady Rutherford's passion for fresh air led to driving sometimes with the hood folded back in weather which any ordinary person would avoid. It was amusing to see them both wrapping up for such a drive in woollens, greatcoats, gloves and goggles, and with hot water bottles on the lap of the driver and at the feet of the rugged-up passenger.[4]

With the Rutherfords' daughter Eileen married to Ralph Fowler and raising her own family, holidays to the Welsh countryside came to comprise only the two Antipodean couples. As a result, Mark and Rosa Oliphant become de facto family for the Rutherfords when the older couple weren't spending time with their grandchildren. And, despite the many mutualities of the two men's professional lives, the relationship was never more effortlessly happy than when away from the laboratory, simply enjoying each other's company.

At Celyn, Mark and Rosa flaunted their comparative youth by tackling steep walking trails affording unhindered panoramas of the escarpments that rose, brooding and bare, above the treeline. These paths also provided spectacular views back down the valley, across the dark, still waters of stream-fed lakes Llyn Dinas and Llyn Gwynant, to the closest village of Beddgelert. On cool days, a thin blanket of white smoke wisped comfortingly across the slate roofs of the village's dark stone homes.

The Rutherfords would restrict themselves to the lower reaches, skirting fields that echoed with bleating sheep, or taking the short walk to the only nearby shop, where essentials, including a weekly bread delivery and locally churned butter, accounted for most of the available stock.

Occasionally, all four would set out on the twelve-kilometre return walk to Beddgelert for more substantial supplies. Rutherford would inevitably give up when the party neared the halfway point at the most distant of Llyn Dinas's rocky shores, and sit and stoke his pipe into life while waiting for the group to reappear. Consequently, that trip was more often completed by car.

Come winter at Celyn, Rutherford rarely found it necessary to venture outdoors at all. Instead, he would settle into a deep lounge chair pulled up to the hearth and pass the days reading forwarded mail, newspapers and books on varied non-fiction topics. His insights into the demise of the region's slate pits and the ultra-narrow gauge railway that had serviced them suggested that some of those texts were local histories.

Oliphant's memories of those calming days would remain as clear as the air that hung heavy with humidity during summer breaks, and numbingly chilled as sunlight hours reduced.

> The stone walls and floors of Celyn seemed always damp, and the bedrooms never warmed. The fine living room was heated with a coal fire, and in the cold and damp weather, Rutherford spent most of his time there. He wore a golfing suit with plus-fours and cloth cap. On a warm day, if he went for a walk, he perspired freely. By the time he demanded a rest, seated on a convenient rock, he was breathless and often irritable.
>
> In spring, foxgloves in every shade of colour fringed the stone walls and rocks, and grass sprouted green once more, and the lambs gambolled about their mothers. In summer, it could be very hot in the valley, but the weather was unpredictable and rain seldom far away. In winter the valley was a cold, lonely place, the grass frozen brown and yellow, and clouds scudding across the mountain tops, the summit of Snowdon barely visible. But it held a strange beauty of its own ...[5]

* * *

Rutherford's fondness for Oliphant might have found its genesis in the commonality of their pedigrees, but it grew readily from there into a meeting of the minds. While other Rutherford acolytes including Moseley, Bohr, Chadwick and Kapitza forged strong personal and professional ties with their leader, the bond he shared with Oliphant trumped them on several levels. While the pair's shared passion for pragmatic science made them kindred spirits within the laboratory, their abiding love of nature born from boyhoods spent outdoors elevated their relationship to another plane.

At the Cavendish, however, they were just two men revelling in their shared fascination for particles.

As a gifted experimentalist, but a physicist with a few self-confessed gaps in knowledge, Oliphant saw that the surest path to success in such a talented pool was to model himself, as much as practicable, on the world's foremost experimental scientist – who, as convenience would have it, was also his supervising professor and personal friend.

Rutherford's inspiration, before coming under J.J. Thomson's influence when he first arrived at the Cavendish, had been the British chemist and physicist Michael Faraday. Whether Rutherford also saw himself in Faraday's back story – apprenticed to a London bookbinder at age fourteen, Faraday was self-taught and was unsuccessful in his early attempts to secure work at the Royal Institution's laboratories – is unclear. However, Rutherford was a voracious reader of history and biography, and became a keen student of Faraday's diaries. These were not mere jottings on daily experiences by the man regarded as science's most gifted benchtop practitioner before Rutherford, but rather detailed experimental notes.

What Rutherford gleaned from Faraday, who died four years before the New Zealander's birth, Oliphant sought from Rutherford. 'Faraday's work laid the foundations of electrical engineering,' Oliphant would explain in his 1946 Rutherford Lecture to Britain's Physical Society. 'Rutherford's is the cornerstone upon which is based the exploitation of atomic energy.'[6] Faraday's rise from a poor family and rudimentary school education to establish himself as perhaps the greatest experimental scientist the world has known was both comforting and compelling to Oliphant too, given his own circumstances.

As Rutherford saw it, the basis of Faraday's genius lay in the simplicity of his methods. That was therefore the fundamental trait that Rutherford came to demand of his own work, and that of his students. In correspondence with acclaimed Japanese physicist Hantaro Nagaoka in 1911, Rutherford had set down the philosophy that he would extol at the Cavendish: 'I have always been a strong believer in attacking scientific problems in the simplest possible way, for I think that a large amount of time is wasted in building up complicated apparatus when a little forethought might have saved much time and much expense.'[7]

In a presentation he delivered years later, Rutherford further distilled that sentiment: 'If you can't explain to the charwoman scrubbing your laboratory floor what you are doing, you don't know what you are doing.'[8]

'Rutherford did not expect spectacular results, though his deep pleasure was evident when they came, but he did expect devotion to research,' Oliphant found as he developed a deeper understanding of his mentor. 'What is more, he was confident that knowledge of nature was as yet elementary, and that research into any of the phenomena of nature could be rewarding. He often likened science to Tom Tiddler's ground [a children's game], for wherever one dug with intelligence and energy, no matter how many times it had been dug over before, something interesting was bound to turn up.'[9]

Rutherford was known to evaluate a scholar's worth solely on their experimental output, and he found little use in people who were not productively working. On one afternoon, as he chaperoned his young grandson around the Cavendish, the boy pointed at a research student hunched over equipment and asked of Rutherford, with a child's innocence, 'What's that man doing, Grandad?' Rutherford peered into the lab's semi-darkness to study the figure before eventually replying, 'I don't know my boy – I've often wondered myself.'[10]

There were some within the Cavendish, notably those of British heritage, who suspected their director's faith in research students like Oliphant who hailed from elsewhere in the Commonwealth – a cohort that, at times, accounted for a third of the laboratory's number – represented something of a blind spot. Rutherford's rationale, however, was that he simply wanted people whose talent for the subject matter was only exceeded by their capacity to get things done. That a high proportion of such researchers came from the Dominions, where opportunity rarely arrived upon a pewter platter, was, to him, happy coincidence.

In addition to the professional and practical guidance he took from Rutherford, Oliphant found in his teacher characteristics shared with his adored father, Baron. Among them were a heartfelt interest in people and a willingness to hear their stories regardless of rank or reputation. Rutherford had an appreciation of basic human decency that Oliphant greatly prized. At the end of any of his public discourses, Rutherford would personally thank every technician who had helped out with experimental equipment or visual aids.

'He loved to tell anyone who would listen about his own work and interests, but he was as ready to discuss cosmology with Eddington as rabbits with his gardener at his country cottage; to listen avidly to the story of the conquest of the tsetse fly in Africa, or to argue holism with Field-Marshal Smuts,' Oliphant later observed.[11]

Like Baron Oliphant, Rutherford was also known as something of a soft touch for charitable causes he felt were credible and worthwhile. His harshly pragmatic wife would often intervene to prevent him from volunteering for events he had neither time nor stamina to attend. During 1925, in his role as president of the prestigious Royal Society, Rutherford had been engaged to deliver twenty speeches throughout Britain, in addition to the six-month lecture tour that took him to Australia and New Zealand. He had turned down a further sixty requests to address crowds in his adopted homeland that year.

The fact that he had become only the second New Zealander (after ornithologist Walter Buller) to be confirmed as a Fellow of the Royal Society was a source of great scientific pride in his homeland. It was an achievement subsequently eclipsed, however, when the Society's Fellows elected Rutherford as the 44th man – and first Antipodean – in more than 260 years to be installed as president for a term strictly mandated at five years. By doing so, Rutherford joined an illustrious lineage of former office-holders who included architect and astronomer Sir Christopher Wren, Sir Isaac Newton, botanist Sir Joseph Banks and Sir Joseph John (J.J.) Thomson. Far from a mere honorary title, the presidency added considerably to his already frenetic workload.

Where he could employ his profile for social good, however, Rutherford regularly found time to engage in a little activism. When the stampede of male students back to the universities after the Great War brought suggestions from campus administrators that the pressure be eased by refusing admission to women, Rutherford co-signed a letter of opposition sent to *The Times*, along with Cambridge's chemistry professor, William Pope. They asserted: 'We welcome the presence of women in our laboratories on the ground that residence in this University is intended to fit the rising generation to take its proper place in the outside world, where ... in the present stage of the world's affairs, we can afford less than ever before to neglect the training and cultivation of all the young intelligence available.'[12]

The role Rutherford played in helping to restock Belgian libraries destroyed in the war of 1914 to 1918, and also the financial aid he mustered to re-establish Vienna's Radium Institute when it remained blacklisted as an 'enemy institution' post-war, were never forgotten among the Continent's science community. If Mark Oliphant's fondness for tackling public causes had not been sufficiently cultivated by his 'do-gooder' family, then the exemplar he found in Rutherford was just as powerful.

'[Rutherford] subscribed wholeheartedly to an opinion expressed by Bertrand Russell, that the world would be an infinitely pleasanter place if men would but learn to seek their own happiness, rather than the misery of others,' Oliphant later wrote, citing another Trinity College figure, one with whom he would later build a close rapport.[13]

Rutherford also liked to poke fun at what he saw as Oliphant's 'Shavian ideas' – reflecting those of playwright George Bernard Shaw, often at odds with the wider community's – as well as his habit of reading the *New Statesman*. It was this magazine, founded by the socialist Fabian Society, that had helped form Baron Oliphant's political views, and which the socially conservative Rutherford jokingly characterised in Mark Oliphant's presence as 'dangerously subversive'.

If Rutherford needed a reciprocal demonstration of Oliphant's generosity of spirit, it arrived one September Sunday in 1930, during one of Lady Mary's afternoon teas. An uninvited visitor arrived at the front door of Newnham Cottage, suitcase in hand, but brandishing no formal invitation to the social gathering within. The young man explained that he had travelled from Eastern Punjab in India with the goal of working with Rutherford at the Cavendish.

Neither Rutherford nor his wife was impressed by such brazenness, but they were placated by Oliphant and Chadwick, who convinced

Rutherford not to send the humble, if misguided, traveller on his way. Oliphant then offered to take on the visitor as a research student, and they worked closely together in researching the artificial disintegration of atomic nuclei.

The knowledge of nuclear physics that the man – who became Professor Rafi Muhammad Chaudhry – gleaned from his time at the Cavendish and from subsequent research work undertaken elsewhere in Britain would prove crucial when he returned to Lahore, in what became Pakistan after post-colonial partition. With the help of Oliphant's overt lobbying, Chaudhry took up a position at Lahore's Government College University, where he would be recognised as a driving force behind Pakistan's nuclear weapons research program.

In his role as Professor of Nuclear Physics, he received ongoing support from Oliphant, who helped him secure equipment grants, and the two men maintained a lifelong correspondence. It was only a force more formidable than Mary Rutherford – the discriminatory White Australia policy – that prevented Oliphant from securing his esteemed colleague a position at Canberra's new national university in the 1950s.

* * *

As Oliphant grew more comfortable and credentialled among the Cavendish hierarchy, and as Rutherford's commitments away from the laboratory mounted, the professor increasingly turned to his Australian protégé to fill in for him at his scheduled lectures on atomic physics.

> On these occasions, he would call me into his office to sit beside him while he explained what I should talk about, with the aid of an extraordinary collection of notes, written on odd scraps of paper and pinned together in the right order. He had used these notes for many years, so that there were copious amendments and additions, always in pencil, some of which had become almost too faint to read. We went through these carefully, sometimes for an hour or more, adding, crossing out, and getting the papers into hopeless disorder.[14]

Unfortunately for Rosa, her husband's increased kudos in Rutherford's eyes did not translate to any significant easing of austerity measures at home. After almost a year at Bateman Street, Mark found the couple slightly more expansive lodgings on Hinton Avenue, where they shared bathroom facilities with the landlady, who lived in a basement flat. The residence proved to be so numbingly cold they would awake on some winter mornings to find the water in the jug on the bedroom washstand frozen solid. It was a clear sign that alternative lodgings were urgently needed.

They soon shifted to a small terrace property in Grantchester Meadows, a quiet street just over a kilometre south of the city's heart. The extra distance from central Cambridge meant that walking to and from the Cavendish Laboratory and shops and services was no longer practical for the Oliphants; it therefore necessitated the purchase of cheap second-hand bicycles. But Mark's improving circumstances also allowed them to buy their first wireless radio set for £1, though it did not include an electrodynamic speaker, so listening was an individual pursuit, conducted through rudimentary headphones.

It was not until the summer of 1928 that Mark gained access to a vehicle, sharing with his Australian colleague, Cecil Eddy, the £50 cost of a well-worn Citroën convertible, complete with celluloid side curtains that flapped at a furious pace, even when the vehicle was moving at one considerably less so. Weekends were then often spent exploring further afield, with the motorised transport opening up parts of the English countryside that could not be feasibly reached by bike.

Around the same time, Oliphant applied for his bursary to be urgently upgraded to a 'senior studentship', which would net him an additional £150 per annum for the next two years. This increase was offered by the 1851 Commission to 'a few selected students of exceptional promise and proved capacity for original work'.

It would be months of anxious waiting before he learned that the Commission had found in his favour, with a note from the secretary adding: 'I am very glad to know that as a married man the remuneration of the senior scholarship enables you to live in greater comfort, as I realise that you must have had a great struggle to get through on your overseas scholarship stipend.'[15] By that time, Oliphant had been at the Cavendish more than two years and was close to completing his PhD.

Having submitted an expansive thesis that detailed his investigation of the ways metal surfaces behaved when bombarded by positive ions, he needed only to pass the oral component of the examination to earn his doctorate. The interview lasted well over an hour, and was conducted by Rutherford and his Cavendish colleague Charles Ellis, who would ultimately vie with Oliphant for the role of laboratory assistant director.

Like Chadwick, Ellis had been in Germany – albeit on holiday – when war broke out in 1914, and was likewise detained at Ruhleben, where he helped Chadwick establish his makeshift laboratory. At conflict's end, he entered Trinity College and was involved in research work at the Cavendish when Rutherford arrived in 1919. He was a pivotal member of the laboratory team, though his work explored the effects of beta and gamma radiation, while Chadwick and Rutherford focused on the possibilities presented by alpha radiation and particles.

Years after he successfully negotiated the examination by Rutherford and Ellis, Oliphant found himself in Ellis's shoes, alongside the professor, in a similar interrogation of a research student, Bertram Vivian (later Baron) Bowden, who was seeking his doctorate qualification. Oliphant learned that asking the questions could be almost as intimidating as having to answer them, after his opening query was met with wide-eyed terror from the interviewee, who eventually spluttered: 'I'm afraid I don't know.'

At which point Rutherford stepped in, turning to his co-inquisitor and harrumphing: 'What's more to the point, Oliphant, neither would you if you hadn't looked it up ten minutes ago. Let's have the next question.'[16]

* * *

Although confirmation of her husband as 'Dr Oliphant' brought both prestige and pecuniary benefits for Rosa, she continued to feel that Mark devoted an unrealistic proportion of his time and energy to the research work that consumed him. His improved circumstances had eased the worst pain of poverty, and also allowed them occasional luxuries. But her sense of isolation lingered.

The plan forged for their Cambridge adventure when the couple left Adelaide – that they would remain in Britain only for the duration of Mark's scholarship then return to Australia – seemed less likely to be enacted as they slowly settled into life as expatriates in England. Despite now holding the doctorate that had been his primary objective when setting out on his initial voyage, Mark could no more consider quitting Cambridge than he might ponder a return to silversmithing. Having secured the financial wherewithal to continue his research work until at least mid-1933, he was now as much a part of Rutherford's vision and the laboratory's future as the Cavendish was embedded in his own soul.

The loneliness that Rosa continued to feel, despite the kindness exhibited by Lady Rutherford and Elizabeth Cockcroft and a widening circle of other friends, became acute when the Oliphants decided to start the family they both so craved, and Rosa suffered a series of devastating miscarriages. The pain and loss she and Mark felt but stoically concealed were compounded by news, in late 1929, that the Cockcrofts' almost three-year-old son, Timothy, had died suddenly from a severe asthma attack. Rosa and Mark noted how John Cockcroft threw himself ever more single-mindedly into his work.

Then, on 6 October 1930, Rosa Oliphant gave birth to a son, Geoffrey, at the couple's Grantchester Meadows home. She had been confined to bed for much of a difficult pregnancy, but five and a half years and 16,000 kilometres from their wedding in Adelaide, Mark and Rosa Oliphant at last had a child of their own. Rosa was attended throughout by their family physician and a Swiss nurse recommended by the Rutherfords.

Fortune was now shining upon Mark Oliphant. Shortly after the arrival of a healthy son, he received news of his first successful grant application – albeit for the modest sum of £80, to cover the cost of apparatus. That was soon dwarfed by news from the Royal Society – doubtless driven by Rutherford, then in the final throes of his five-year president's term – that Oliphant had been appointed a Messel Fellow. The fellowship was funded from the estate of German-born industrial chemist Dr Rudolph Messel, who died in his adopted home city of

London and bequeathed a vast portion of his personal fortune to the Royal Society, of which he was a fellow. This honour increased Oliphant's annual income by a further £150 a year, with an accompanying pledge for an annual allowance of £600 (around £40,000 today) for two years from 1932, when his senior studentship expired.

In the two years since Oliphant's finances were so grim he had been compelled to write, cap in hand, to the 1851 Commission's secretariat in London, chasing the slim chance of a scholarship upgrade, his yearly income had more than doubled. To celebrate, and to make use of the return fare – valid for five years – that was part of his 1851 scholarship arrangement, he booked passage in late June 1931 for Rosa and Geoffrey to accompany him on a triumphant visit to Adelaide. It was almost four years since they had sailed for Cambridge, and both their families – anxious to see the couple and their blond-haired, now ten-month-old boy – planned to give them rollicking welcomes when they set foot on South Australian soil following a six-week sea voyage.

Weeks after arriving, amid much fêting of Mark's academic achievements and joy at the new family addition, Oliphant announced that he must scurry back to Cambridge. The summer dormancy of June to September would soon end, and the university would re-engage for Michaelmas term. He advised Rosa, however, to stay on in Adelaide and spend additional time with her family, thereby not subjecting Geoffrey to England's approaching winter, and its damp threat to infant children.

His wife was uncomfortable with the prospect of a prolonged separation, but she stood little hope of forcing a backdown by her strong-willed husband. It would be six months before all three were reunited, as 1932 slowly embraced the warmth of the northern spring.

* * *

Around the same time as the Oliphants were cautiously welcoming news that Rosa was pregnant with Geoffrey, Rutherford learned he was to become a grandfather for the fourth time. In the spring of 1930, Eileen Fowler – not yet thirty – revealed that she and Ralph were expecting another child.

Not unlike Rosa Oliphant, Eileen was confined to bed for all but the first month of her pregnancy. She had been warned against having more children due to health issues that had become increasingly serious during earlier deliveries, including the development of a tubercular spot on one of her lungs. However, her keenness to add to a family that already included sons Peter (then aged seven) and Patrick (three) and daughter Elizabeth (known as Liddy, aged five) meant she was prepared to shoulder that risk.

Although Oliphant sensed Ernest and Mary Rutherford's misgivings as the pregnancy progressed, Eileen remained unfazed by the medical warnings. Mary accordingly went ahead with plans to return to New Zealand over the Christmas and New Year of 1930 to 1931 to visit her own ageing mother in Christchurch. 'I am sorry to be away when her baby arrives in December but am not at all worried about her,' Mary wrote to her mother-in-law, confirming her imminent homecoming. 'Eileen looks very peaky and we think she shouldn't have thought of having a fourth, but she was keen to.'[17]

The child – a daughter, Ruth – was born on 14 December 1930 without incident and, despite a bout of gastric flu that ran through the household, there was no indication of pending trouble. That was until nine days later when, as Ernest dressed to attend Tuesday-night dinner at Trinity College's Great Hall, a nurse arrived from the Fowler house to tell him that Eileen, his only child, had died due to a pulmonary embolism – a clot that had entered her bloodstream and led to cardiac arrest. It was, at that time, a not uncommon consequence of childbirth.

Rutherford drove immediately to the Fowlers' house, and next day – Christmas Eve – sent a telegram to Mary via her brother, who was hosting the Newton family's Christmas dinner in Christchurch.

'Eileen died suddenly but peacefully Wed evening. Embolism,' he wired. 'Baby well … all well. No need to change your programme. Rutherford.'[18]

The news arrived just hours before Mary's family was to gather for dinner on Christmas Eve. Charlie Newton chose not to pass on the information until after they had eaten.

While the private grief elicited in Ernest and Mary by the loss of their sole child can only be imagined, outwardly their response was one of fatalistic acceptance, given Eileen's known health risks.

In the letters they penned each other from opposite hemispheres that bleak Christmas Day, Ernest wrote: 'It is a sad end for Eileen's adventure but it may be called in a sense a happy end for I was always afraid of her becoming an invalid. If her lung trouble had flashed out again it would have been a bad complication in any case.'[19] Mary echoed the same sentiment in her response: 'Can't help feeling that poor Eileen has been saved a life of invalidism.'[20]

So as not to spoil their Christmas Day, the Fowler children were told their mother was simply too unwell to join them for Yuletide celebrations. It prompted the eldest, Peter, to inquire: 'Is she ever going to get better?'[21] The terrible truth was broken to them the next morning.

Mary then remained in New Zealand for several months, until the British winter had fully passed. Perhaps as a means of embracing some measure of normality to help quell their collective grief, Mary wrote to Ernest during the early weeks of 1931 recounting mundane details, including updates on the success of the new year diet she had adopted.

With no immediate family to share or alleviate his sorrow, Rutherford withdrew – at Mary's suggestion – to the untamed, snow-capped vastness of North Wales. To nourish his being, he took with him the couple's housekeeper Eileen 'Bay' de Renzie, Mary's cousin, who was not much older than their own departed Eileen. To repair his soul, he asked Mark Oliphant to join him. The Australian left his wife and three-month-old son in Cambridge to support his anguished friend.

Oliphant spent a week with Rutherford at Celyn in the bleak early days of 1931, while Miss de Renzie lavished the professor with care and affection. To Oliphant, he seemed suddenly older than his fifty-nine years, and exhibited telltale signs of the toll extracted. His handwriting appeared less legible, and he tired more readily than on previous excursions into the valley.[22]

* * *

The desolation brought by Eileen's death was also painfully difficult to reconcile with the celebratory news that Rutherford was compelled to keep secret until after Christmas. In the weeks before the tragedy, he had received confidential advice that his name had been submitted to King George V for bestowal of 'the dignity of a Baronetcy of the United Kingdom' in the 1931 New Year's Honours. As a life peer, he would not only be entitled to a seat in Britain's House of Lords, but could also nominate his own territorial adjunct, to distinguish himself from an earlier Baron Rutherford, whose lineage had long since expired.

Due to strict protocols surrounding release of this information, Rutherford had felt unable to reveal the honour to Eileen before her death. However, he had penned a letter to his wife in New Zealand dated 19 December, in the knowledge that the mail steamer would only dock at Christchurch safely after the list was published.

Mary wrote back, counselling that the location appended to his title should be readily identifiable with the nation of their birth, not simply as 'Lord Rutherford of Canterbury' or 'Christchurch' which would potentially be interpreted in the northern hemisphere as an English reference. 'Havelock, where you got your first educational start seems the most appropriate and sounds well too. Lord Rutherford of Havelock ...'[23]

Whether it was boldness born of separation, or mere personal preference, Ernest overlooked his wife's suggestion and opted instead for 'Lord Rutherford of Nelson'. In addition to the kiwi and the Maori warrior that adorned the coat of arms he had drafted, Rutherford included Hermes Trismegistus, whose ancient writings gave birth to alchemy, and inscribed it with the motto *'Primordia quaerere rerum'*, which Oliphant would doubtless have identified from his schoolboy Latin as 'To seek first principles'.

Unlike Rutherford's sponsor for the honour, fellow physicist Robert Strutt – whose father had served as the second Cavendish Professor at Cambridge, after the family adopted the title Rayleigh when bestowed with a peerage – Rutherford chose to retain his surname, as he explained in his letter to Mary:

> So I suppose if it goes through I shall be styled Lord Rutherford but there will be no change in your title. It has been rather a worrying business, for I was very uncertain whether to accept; for a title of this sort is of little use to people like ourselves with no social ambitions. I was pressed by Lord Parmoor [Lord President of the Privy Council] – through whose department it goes – that I ought to accept on general grounds as a recognition of the importance of science to the nation ... Of course, I do not intend [it] to make any difference in our mode of life.[24]

The telegram he sent to his widowed mother in New Plymouth – his father having died two years earlier, aged eighty-nine – was more succinct: 'Now Lord Rutherford. More your honour than mine.'[25]

The humility he displayed was a quality clearly inherited from his mother. When New Zealand's governor-general, Lord Bledisloe, gushed with national pride at the news and declared to Martha Rutherford that 'you must be very proud of your illustrious son', she responded coolly: 'No more than I am the rest of my sons.'[26]

This personal acclaim sat jarringly alongside the pain of bereavement through those bitingly bitter weeks at Celyn. However, while Oliphant noted the melancholy that pervaded the damp, dark house in the early part of that sad stay, he also detected a gradual lightening of Rutherford's mood as time passed.

> He pretended that the new honour was of secondary importance to him, but he could not help showing how he revelled in this recognition of his standing as a man of science. Despite his sadness, one morning he could be heard singing as he dressed. When the heaps of letters of congratulations, sent on from Cambridge, arrived each morning, he slit the envelopes with obvious pleasure, passing the contents to Miss de Renzie with some bright quip about the sender. Some he answered at once, but others were kept till he returned to Cambridge.[27]

* * *

The shadow of his daughter's death would forever remain with Rutherford, but gradually the letters he wrote to Mary bore a more familiar theme, as he documented the latest focus of his laboratory work. The single-minded intent that had driven his earlier discoveries was beginning to reappear. While the dawning of 1931 had brought with it darkness and despair, the arrival of 1932 presented a markedly different outlook.

It was a year that would be retrospectively hailed as nuclear physics' *annus mirabilis*.

10

THE GOLDEN YEAR

Cambridge, 1932

As leader of the first assault on the tightly bound nuclei of light elements such as hydrogen and nitrogen, Rutherford understood more keenly than many that heavy artillery was needed to bombard the atoms at the far end of the periodic table. These materials, culminating in uranium – then the heaviest known naturally occurring element – contained such strong electrical forces of repulsion at their core that they were immune to the peppering of Rutherford's beloved alpha particle bullets.

In his 1927 address as Royal Society President, Rutherford had speculated that new tools capable of producing particles that moved in greater quantities and with increased energy were needed if nuclear physics were to continue progressing at the rate it had evolved over the past two decades.

> It has long been my ambition to have available for study a copious supply of atoms and electrons that have an individual energy far transcending that of the alpha and beta particles from radioactive bodies. I am hopeful that I may yet have my wish fulfilled, but it is obvious that many experimental difficulties will have to be surmounted before this can be realised, even on a laboratory scale.[1]

A major step towards overcoming that obstacle was taken just over a year later, in January 1929, when Russian theoretical physicist George Gamow visited the Cavendish. Employing a new model utilising the recently postulated wave mechanics theory – a model formulated in conjunction with Niels Bohr at his Copenhagen Institute – Gamow proposed a fresh notion of how particles might penetrate the atomic nucleus. His theory was subsequently taken up by Oliphant's close friend John Cockcroft, who saw that it raised the prospect of fully disintegrating nuclei of light atoms by bombarding them with artificially accelerated protons.

Much as Rutherford distrusted the type of complex theory that Gamow and others expounded, and disliked even more a reliance on expensive machinery to solve problems at the expense of table-top experiments and cognitive ingenuity, the Cavendish nevertheless needed to embrace progress. Starting with the massive electromagnetic fields that Kapitza generated in his dedicated laboratory space, the string-and-sealing-wax model began to make way for electrical wiring and metal welds.

In taking up Rutherford's challenge to generate faster, more powerful particle bullets, Cockcroft joined forces with Ernest Walton, the Irish scholarship holder who had arrived on that same October morning as Oliphant in 1927. The premise of the pair's work was to generate an abundant supply of protons by discharging an electrical current through hydrogen, and then using a strong electrical force to direct them.

It was an aspiration of enormous complexity, and one that would require equally intricate equipment. But, if successful, it would represent a high watermark for the Cavendish on the rapidly changing landscape of nuclear physics.

At least it would have done, had another seismic discovery not been made at the laboratory early in 1932, a few months before Cockcroft and Walton's project reached fruition.

* * *

Rutherford had first aired his belief that the atomic nucleus might, in addition to charge-carrying protons, house an as yet unidentified neutral particle, in the section of his 1920 Bakerian Lecture that drew little attention. He had speculated that this 'neutral doublet', as he called it, would exhibit unique properties, among them a capacity to move freely through matter. It would therefore – in Rutherford's futuristic vision – be readily able to penetrate the defences of the most intensely fortified heavy elements, because its neutral charge would mean it was immune to electrostatic repulsion. In modern military parlance, it would be a sub-atomic stealth bomber.

If the broader scientific community, used to dealing only in proven experimental outcomes, gave little heed to the theory, and Rutherford himself rarely revisited it in public as he busied himself with delivering more immediate outcomes, the comments had resonated with James Chadwick. He had been among the audience at that 1920 lecture to the Royal Society in London. Quietly at first, he had begun his hunt for the elusive neutron that would dramatically alter the world's understanding of matter, and the very future of humankind.

Across a decade or more, Chadwick – often in collaboration with Rutherford, and amid his myriad supervisory and administrative duties at the Cavendish – had patiently performed one experiment after another. They involved searching for evidence of highly penetrative gamma radiation spontaneously produced by hydrogen, then by any number of rare gases, and then an assortment of rare elements.

None of them had yielded worthwhile results. 'I did quite a number of silly experiments,' Chadwick later wrote, adding: 'I must say, the silliest were done by Rutherford.'[2]

In the course of this work, Chadwick had become intrigued by the results achieved when the alkaline earth metal beryllium was bombarded with radioactive particles. These, in turn, had ejected particles of high energy and penetration. 'For a short but exciting time we thought we had found some evidence of the neutron,' Chadwick later reported. 'But somehow the evidence faded away. I was still groping in the dark.'[3]

Then, in the latter half of 1931, Chadwick had been alerted to results published in Paris by Marie Curie's scientist daughter, Irène, and her husband Frédéric Joliot-Curie. They reported observations of paraffin wax that emitted high-velocity protons when bombarded by gamma radiation.

However, their conclusion that the highly penetrative gamma rays were causing protons to be fired from the paraffin at speed did not make sense to Chadwick. Gamma rays were known to be capable of knocking out electrons, but a proton was almost 2000 times heavier than an electron, and similarly difficult to budge. It was as if a table tennis ball fired at a bowling ball had been causing the latter to hurtle away. Chadwick suspected that the French experiment might, albeit unknowingly, have revealed the true energy source to be the neutral particle.

In keeping with his quiet, unassuming character, Chadwick waited an hour or more for his daily 11am meeting with Rutherford before explaining to him the details of what he had read in the *Comptes Rendus* physics journal, and what he understood it to signify.

'As I told him about the Curie-Joliot observation and their views on it,' Chadwick recounted of that exchange with Rutherford, 'I saw his growing amazement; and finally he burst out "I don't believe it". Such an impatient remark was utterly out of character, and in all my long association with him I recall no similar occasion.'[4]

Using polonium – the element identified by Marie Curie, and named after her homeland – as his radiation source, Chadwick found that the alpha particles fired at beryllium produced rays that were then trained upon a paraffin target. It was then found these rays could pass unimpeded through a sheet of lead up to two centimetres thick. This could only occur if those particles held a similar substantial mass to a proton but carried no charge, which would have seen them electrically repelled. It was proof that the neutral particle existed.

Chadwick spent the next ten days, surviving on two or three hours' sleep each night, substantiating his contention.

The existence of a third essential constituent of matter, alongside the proton and electron, was formally announced to the world in a letter published in *Nature* on 27 February 1932, entitled 'Possible Existence of a Neutron'.

There was no equivocation such as he had shown to Rutherford when, fortified by dinner and drinks at Trinity Hall beforehand, Chadwick confirmed his stunning revelation to the Cavendish's Kapitza Club. He then pronounced: 'Now I want to be chloroformed and put to bed for a fortnight.'[5]

Oliphant would never forget the exhilaration felt by the entire Cavendish team on that occasion.

> The intense excitement of all in the Cavendish, including Rutherford, was already remarkable, for we had heard rumours of Chadwick's results. His account of the experiments was extremely lucid and convincing, and the ovation he received from his audience was spontaneous and warm. All enjoyed the story of a long quest, carried through with such persistence and vision, and they rejoiced in the success of a colleague.[6]

On the strength of his discovery, Chadwick would receive the physics Nobel Prize in 1935, thus unleashing a stampede of experimentation that ultimately yielded an atomic bomb.

In accepting that he and his wife had achieved the same results, yet failed to grasp their place in scientific history, Frédéric Joliot-Curie wrote:

> The word neutron had been used by the genius Rutherford … at a conference to denote a hypothetical neutral particle which, together with protons, made up the nucleus. This hypothesis had escaped the attention of most physicists, including ourselves. But it was still present at the Cavendish where Chadwick worked … Old laboratories with long tradition always have hidden riches.[7]

While the identification of a new particle was of front-page importance to the world at large, its significance in the field of atomic research was, quite literally, earth shattering. In scale, the new particle was similar to a proton, but its absence of charge meant that the electrical barrier blocking intruders from entering an atom's core – particularly in elements such as uranium with the greatest and most impenetrable atomic mass – was no obstacle despite the neutron's size.

'In fact,' Nobel Prize winning physicist I.I. (Isidor) Rabi later declared, 'the forces of attraction which hold nuclei together may well pull the neutron into the nucleus. When a neutron enters a nucleus, the effects are about as catastrophic as if the moon struck the earth.'[8]

* * *

While Chadwick's triumph was toasted by the laboratory, the work of Cockcroft and Walton had attracted little notice. In their efforts to verify Rutherford's other hypothesis, they were still attempting to generate sufficiently large electrical currents to enable protons to be accelerated to similarly high energies. This would allow them to be employed as projectiles to disintegrate atomic nuclei.

If Rutherford had begun his investigation of sub-atomic structure by firing alpha particle bullets from a lead-lined box, particle accelerators were seen as the field gun, capable of smashing apart the most intractable forms of matter. But at the start of 1932, as Rutherford had suggested five years earlier, that point appeared dauntingly distant.

Nor was the quest to achieve higher voltages in the name of science confined to the Cavendish. Years before Cockcroft and Walton began assembling a rectifying circuit from items they sourced and scrounged, three members of the University of Berlin's Physics Institute had taken their ambitious project to the Italian Alps. They had installed an antenna between two mountains to await the regular summer thunderstorms, and when they came, saw sparks jumping almost twenty metres between metal spheres strung on lengths of steel cable. The resulting charge was estimated to deliver an electrical potential of around 15 million volts. It also cost the life of one of the researchers, who was blown off the mountainside by the unforgiving force they sought to tame.

Attempts to artificially replicate massive voltages at Cambridge also brought a very real risk. Oliphant would ruefully recall the day he went to make an adjustment to electrical apparatus that was, in the interests of safety, located behind one of the lab's internal brick walls. However, on this occasion, his fellow experimenter Bernard Kinsey had failed to switch off the 20,000-volt source beforehand. So when Oliphant – standing on the basement's stone floor, which was forever slightly damp like the jute sacks of his boyhood pranks – reached for the equipment, he was sent flying by a shock that rendered him unconscious.

He woke to find smoke wafting from the molten soles of his shoes, and Kinsey bent over him. 'My God, my God, what have I done, have I killed him?' Kinsey fretted.

At which point Rutherford appeared, railing at Kinsey, whose dread had eased when Oliphant came to.

Kinsey then launched his defence: 'But, God's bladders Professor, how was I to know he'd touch the bloody thing?'[9]

Four decades later, Oliphant would receive a letter from another former Cavendish colleague who recalled the burned rubber singed into the 'garage' floor, noting: 'I always thought that those heel marks should have been preserved under a glass plate to show the toughness of the real experimental physicist.'[10]

The perilous pursuit of ever-higher energy sources was also challenging Cockcroft and Walton. They had learned through similarly fraught experience that results gained from a linear accelerator – where electrical fields were used to increase the speed at which charged particles travelled – were limited by the levels of voltage that could be applied to them. Once the electrical energy being imparted reached too great a level, the basic apparatus underwent electrical breakdown, in which the materials employed as insulators began operating as conductors.

As so often manifests along discovery's uncharted path, a solution arose from timely coincidence. In the midst of their so far unsuccessful attempts to accelerate particles, the pair were required to vacate their experimental workspace in the Cavendish basement, so it could be taken over by physical chemists. As a consequence, they built their equipment anew in a high-roofed, reclaimed lecture theatre in the nearby Balfour Building. This allowed Cockcroft to integrate, from the outset, an ingenious voltage-multiplying circuit he had yet to test with higher loads. Just as influential was his successful request for Rutherford to invest £1000 (around £65,000 today) in a 300-kilovolt transformer, a machine capable of delivering similar voltage to an average lightning strike.[11]

It had taken several years for Cockcroft and Walton to piece together the prototype electrostatic accelerator that finally took shape next door to the Cavendish. In its earlier iterations, it had been variously constructed from glass cylinders recycled from petrol pumps, steel bicycle tubing, lead batteries, tungsten wire filaments and sheets of galvanised iron – all made airtight by a specially devised, low-pressure plasticine compound. Once installed in its new premises, the newly improved system could be exhaustively tested with steadily increased voltages. Air leaks caused by softening of the plasticine, or by errant sparks that pierced holes in the glass tubes, were constantly being mended by the two researchers, who often teetered on wooden ladders to reach the extremities of the sprawling device.

Rutherford's patience had begun to fray when he learned that the voltage issues had been resolved but problems remained with the velocity and direction of the proton beam that would carry the atom-smashing bullets. He urged the pair to 'get on with it'; while anxious to see the final effects, he was little interested in the technical specifications required to reach that point.

By mid-spring of 1932, the apparatus underwent its final conditioning, and the positively charged protons, directed by magnetic force, were unleashed upon a lithium target. Then, on 14 April, Walton recorded the telltale incandescence of alpha particles striking a zinc sulfide screen and immediately summoned Cockcroft. Upon verifying the authenticity of his colleague's observations, Cockcroft phoned Rutherford, who arrived shortly afterwards.

Scottish science journalist Ritchie Calder was coincidentally visiting the Cavendish at that epochal moment.

> Ernest, Lord Rutherford of Nelson, pushed aside a Geiger counter, a soldering iron and a cluttering of bits and pieces, and hoisted his six-foot frame and its matching bulk on to the laboratory bench. With his hat tipped to the back of his head and his feet dangling, he might have passed as a farmer at a cattle roup [auction] in his ancestral Perthshire or at a flax-sale in his native South Island, New Zealand. 'Take over Cockcroft,' he said, 'it's your show.'
>
> In the darkened hall, switches were thrown. The generators warmed, with the hum of a gathering storm. There was the throb

of the pumps as they sucked the air out of the vacuum tubes. Lightning crackled and flashed as the high-tension spheres sparked. A tall glass pillar glowed with a luminous blue haze. Presently, there was a clicking sound, and a counter, like a mileage recorder in a motor-car, began to clock in the fragments of the splitting atoms.[12]

Rutherford somehow folded his ample frame into the observation hut, which was not much more than a tea chest fitted with a blackout curtain, to witness the scintillations happening above his head, where the accelerating tube met the screen. After being helped out of the tiny box, he straightened himself upon a nearby stool and proclaimed: 'Those scintillations look mighty like alpha-particle ones. I should know an alpha-particle scintillation when I see one, for I was in at the birth of the alpha particle and I have been observing them ever since.'[13]

Rutherford could now also claim to have been present at the artificial conception of atomic disintegration. The method he had pioneered with natural radiation to chip off sections of atomic matter had been modernised and amplified to fully disintegrate atoms through human means. All that was needed was for Cockcroft and Walton to ascertain the scientific certainty of their breakthrough, while keeping the influx of well-wishers and the morbidly curious at bay.

Working late into the subsequent evenings, the pair studied the results of more collisions and confirmed that their accelerated proton beam had transmuted the light metal lithium's atoms, each with three protons in its nucleus, into beryllium, which has four protons. This new element existed but for fractions of a second before it disintegrated into a pair of nuclear fragments, both containing a helium nucleus. As Rutherford well knew, alpha particles are ionised helium atoms. It was these helium atoms that caused the scintillations they witnessed, their impact with the screen carrying far greater energy than that of the bombarding protons that had set them in motion.

The researchers knew the precise energies of the accelerated protons, as well as the atomic mass of both the lithium and the helium atoms. By measuring the decrease in mass that occurred when the transmutation occurred, and by observing the behaviour of the particles striking the scintillation screen, they calculated the kinetic energy liberated through this process to be 17.2 million electron volts. This number corresponded with Einstein's equation linking mass with energy: the first time $E = mc^2$ had been definitively proven through experiment.

The 'big science' era of high-energy particle accelerators had arrived, at the expense of Rutherford's beloved benchtop experiments. But in applying the same principle he attached to laboratory apparatus, the professor conceded that the only way to fully understand the inner workings of the atom was to take it to pieces.

* * *

In the space of several months, not only had researchers at the Cavendish discovered the neutral particle that would allow far deeper penetration of the atoms of heavy elements, but they had also proven the methodology that would ultimately deliver it. The vast stores of untapped energy within the very essence of matter appeared ripe for harvest, if a means could be found to sustain and contain their release. German Hans Bethe, the 1967 Nobel laureate for physics, considered that everything learned before 1932 comprised 'the pre-history of nuclear physics, and from 1932 on [was] the history of nuclear physics'.[14]

There was now no doubt, however, as to the Cavendish's pre-eminence in the field of experimental physics. Under Rutherford's stewardship, its resolute focus on questions pertaining to the structure and secrets of the atomic nucleus was unprecedented. As Chadwick later noted: 'This was perhaps the first time that a great laboratory had concentrated so large a part of its effort on one particular problem.'[15]

* * *

The photograph of Cavendish Laboratory's class of 1932 depicts nine eventual Nobel Prize recipients, eight of them seated in the front row. Immediately behind one of those – the grand old man J.J. Thomson – stands Mark Oliphant, his right ear tilted tellingly towards the camera.

As was the case for Rutherford upon landing at the Cavendish in 1895 – just as the discovery of x-rays and radiation altered the direction of physics research forever – Oliphant knew he now stood squarely at the epicentre of historic events, which were unfolding in a hurry at a monumental time. He just needed to find the appropriate vehicle to carry him to the heart of the action.

Rutherford's own research path had been redefined more than thirty years earlier when Thomson recognised that the young man would be best deployed in pursuing the big questions that occupied the foremost minds. Rutherford would perpetuate that precedent, by setting his surrogate son onto a problem integral to science's future trajectory – in the process, transforming Oliphant from talented experimentalist to nuclear physicist.

11

FUSION

North Wales and Cambridge, 1932 to 1933

It was early spring of that auspicious year, as the foxgloves blazed purple and pink against Celyn's whitewashed walls, and the air warmed with staccato squawks from newborn lambs, that Rutherford opened the door for Mark Oliphant to join the atom smashers.

Until that holiday weekend – filled with morning walks along the Watkin Path and beside streams of rushing snow-melt overhung by willows' canopies, as well as lazy afternoons surveying the valley from Celyn's terrace – Oliphant's work had centred on the radioactivity of potassium, and the acceleration of low-energy beams of positive ions through gas discharges. Now he would be let loose on the heavy weaponry.

Rutherford mistrusted the new, ever-bigger laboratory machinery, but he understood its necessity – particularly if Cockcroft and Walton's apparatus could be refined to deliver a greater concentration of protons within beams aimed at various metal targets. So, as the Rutherfords and Oliphants enjoyed a weekend in Snowdonia, Rutherford suggested that his most gifted experimental engineer join him in formal collaboration to further the artificial disintegration of nuclei.

Oliphant scarcely needed to mull over such an offer. This was, after all, what he had envisaged when first he set his sights on working with the world's foremost physicist.

The brief, as it was explained during those lazy couple of days at Celyn, was fairly fluid. Oliphant would utilise his flair for design and methodology to improve Cockcroft and Walton's accelerator, in the hope of gaining more accurate measurements of the energies released in the transformation process. Rutherford would then bring to bear his innate understanding of sub-atomic structures, and the implications of their interplay, to interpret those findings. The professor's faith in his protégé was underscored when Oliphant learned he had a free hand to design and execute the research.

Back at Cambridge, Oliphant set himself up in the cool, claustrophobic, stone-floored basement of the Cavendish's renowned Rayleigh Wing. In the adjoining room, a brass plaque denoted the place where Lord Rayleigh had determined the standard unit of electrical measurement subsequently announced as the ohm. That was the same experimental space where J.J. Thomson had discovered the electron thirty-six years earlier, and where Rutherford and Chadwick had then achieved artificial transmutation of elements for the first time.

However, while Oliphant's commission came with Rutherford's full blessing, it was also subject to his intractable budgetary restrictions. The canal-ray (anode-ray) tube through which the protons would hurtle was fitted horizontally, in contrast to the lofty vertical set-up of the Cockcroft–Walton device, because the basement's low ceiling would not accommodate an alternative configuration. Each of the linear amplifier's six segments was housed within a separate biscuit tin. The design was a triumph of oddments made fit for purpose, with seals rendered airtight through strategic placement of plasticine, and the transformers that generated the required voltages mostly salvaged from discarded x-ray machines.

'Like all Cavendish equipment up to that time,' Oliphant later recalled, 'ours was hastily assembled from whatever bits and pieces were available, so that it often gave trouble. Rutherford was very irritated by delays of this kind, but was singularly uninterested in finding the money to buy more reliable components.'[1]

By devising a machine that could fire a greater concentration of protons within a narrower beam, Oliphant was able to operate at less than a third of the voltage generated by Cockcroft and Walton. Had he replicated those earlier currents, it would have been his mortal soul, rather than the soles of his shoes, seared into the Cavendish's stone floors, and its legend, when he strayed too close to the generator.

Not that anyone involved blithely dismissed the danger of their endeavours. In the new accelerator's early iteration, its motor-driven generator and accompanying rectifiers, transformers and condensers were constructed atop a bare pinewood table. The table's legs were strategically placed within oil-filled jars, to negate the electricity's earthward charge. This ploy was revised after the table heated so quickly it caught fire, when the voltage had reached barely half its required output.

However, the breakthrough that allowed Oliphant to make real progress was made not in the acknowledged heart of physics research at Cambridge, but amid the rapidly growing research capabilities across the Atlantic Ocean.

* * *

It was during Rutherford's now-famous 1920 Bakerian Lecture, when he had so presciently floated his theory of the neutron, that he had also wondered aloud about the prospects of a heavy hydrogen atom. This, he foresaw, would be formed with a nuclear mass of two units – a proton and a neutral particle – rather than simply the one, and accompanied by a single electron.

A dozen years later, in that same *annus mirabilis* of 1932, Harold Urey confirmed the theory in his laboratory at Columbia University in New York. He named the isotope 'deuterium' and its atypical nucleus a 'deuteron'.

Soon after this, Urey recognised that it might be possible to concentrate deuterium through the electrolysis of water. This rigorous process was successfully attempted the following year, and the resulting molecule dubbed 'heavy water'. This substance exists naturally, with around one in every 6000 water molecules carrying the additional neutron that lends it double the mass of everyday H_2O. Within a decade, its role as a neutron moderator used to slow the runaway process of uranium fission would make it integral to atomic energy production. But to artificially produce this elusive substance, American chemist G.N. Lewis constructed a twenty-metre-tall separation plant in his laboratory at Berkeley near San Francisco. At peak operation, it gave up less than a teaspoon of heavy water per day. Given the huge amount of energy required to achieve that minimal result, it was deemed useful for little other than experimental value.

Rutherford, however, quickly saw the benefits that might be gained from using these deuterons in disintegration experiments, given that they consisted of a proton as well as the neutron he had already employed to circumvent the electrical forces of atomic nuclei. When Lewis visited the Cavendish in the summer of 1933, he presented Rutherford with a minute pair of sealed glass ampules, inside each of which a single drop of heavy water glistened.

Rutherford entrusted them directly to Oliphant, who, once he had successfully extracted the deuterium to release the prized hydrogen atoms, saw his experimental work gather fresh momentum, and momentous significance.

'Straight away, a new world was opened to us,' Oliphant would recall of the heavy hydrogen's effect on his experimentation. 'There were a new set of explosions, atomic explosions, which were terrific in their intensity, and in the number that took place. It was like entering a new realm of star-watching ... when looking at these scintillations.'[2]

Trained in a beam and mixed with helium, these particles proved highly efficient in disintegrating the nuclei of light metals, but produced some puzzling results. Regardless of the target material into which the beam was fired, the outcome was curiously consistent. The new electronic counting mechanism, which had replaced zinc sulfide screens, detected an array of debris – protons and neutrons – all carrying similar amounts of energy. And if the beam was maintained for a long period, the emanations were increased rather than exhausted.

Finally, with Rutherford's input, Oliphant deduced that whatever element was employed as the target for the accelerated particles was *itself* being covered with a coating of heavy hydrogen. Therefore, the scintillations being recorded came from deuterons bombarding deuterium.

To corroborate this thesis, Oliphant's colleague Paul Harteck, an Austrian physical chemist, prepared small quantities of compounds containing heavy hydrogen, which were then used as targets. The first of these was 'heavy' ammonium chloride, effectively a form of smelling salts. When the beams containing deuterons struck the deuterons embedded in the targets, the pattern from the earlier collisions was repeated as long-range particles were liberated by violent collisions.

Identifying the matter being spat out by this disintegration required detailed measurement of the energy that each emanation carried. This was achieved by calculating the distance they could travel through a range of impossibly thin mica screens, some so sheer they replicated the stopping capability of less than one centimetre of air.

Using these tools, Oliphant found that one group of particles produced by the disintegration could barely make it beyond the mica screen and into the electronic counting chamber. These short-range particles were found to be ions of hydrogen that carried a single charge but also weighed three times as much as ordinary hydrogen.

What Oliphant had artificially produced was a previously unknown radioactive isotope of hydrogen that appears in minute quantities in water; in proportion to natural hydrogen, its presence is $1:10^{-18}$. Its increased mass was the result of a nucleus that contained a proton plus a *pair* of neutrons. Because it was triple the weight and structure of ordinary hydrogen, Rutherford, Oliphant and Harteck christened the new entity 'tritium', and the particles they had discovered 'tritons'.

The finding meant there were now three known types of hydrogen, the universe's most abundant element: ordinary hydrogen with a solitary proton; 'heavy' hydrogen, which also carried a deuteron; and the new tritium, or hydrogen-3. Closer examination revealed that the transformation came about when the high-speed collision of deuterons with deuterium liberated large numbers of protons and neutrons, as well as the proton–double neutron combination at the heart of hydrogen-3.

With one puzzle solved, others immediately emerged – among them, the identity of other particles released in this explosion of sub-atomic fury. That hunt called for technical wizardry that surpassed even Oliphant's.

George Crowe, who served as Rutherford's laboratory steward for so long he lost the tips of several fingers due to the highly radioactive alpha particle source elements he regularly handled, was able to split a mica screen of such gossamer fragility it exerted stopping power the equivalent of just 1.5 millimetres of air. It was only by being fixed to a fine mesh of brass that this barely there filter was able to withstand the atmospheric pressure.

The delicate screen allowed Oliphant to study low-range particles that flew a very short distance from the target. This in turn brought another discovery – in circumstances that Oliphant would delightedly recount to scientific conferences, at public addresses, in media interviews and during fireside chats for the next half a century.

> We found a group of particles which clearly carried a double charge and appeared to be alpha particles, in numbers equal to the protons and tritons. The observation produced consternation among us. The equality of fluxes suggested that all three groups of charged particles originated in the same process. Rutherford produced hypothesis after hypothesis, going back to the records again and again, and doing abortive arithmetic throughout the afternoon. Finally, we gave up and went home to think about it.
>
> I went over all the afternoon's work again, telephoned Cockcroft who had no new ideas to offer, and went to bed tired out. At 3am the telephone rang. Fearing bad news, for a call at that time is always ominous, my wife, who wakens instantly, answered it and came back to tell me that 'the Professor' [Rutherford] wanted to speak to me.
>
> Still drugged with sleep, I hear an apologetic voice express sorrow for waking me, then excitedly say: 'I've got it. Those short-range particles are helium mass three.' Shocked into attention, I asked on what possible grounds could he conclude that this was so, as no possible combination of twice two could give particles of mass three and one of mass unity.
>
> Rutherford roared: 'Reasons! Reasons! I feel it in my water!' He then told me that he believed the helium particle of mass three to

be the companion of a neutron, produced in an alternative reaction which just happened to occur with the same probability as the reaction producing protons and tritons.

I went back to bed, but not to sleep. I called in to see Rutherford at Newnham Cottage after breakfast, and went through his approximate calculations with him. We agreed that the way to clinch the conclusion was to measure, as accurately as we could, the range of the doubly charged group of particles, and the energy of the neutrons … Of course, Rutherford was right.

By the end of the morning we had satisfied ourselves that an alternative reaction of two deuterons, produced a neutron and a helium particle of mass three, the energy released being close to that in the other reaction. The mass of helium three worked out to be a little less than that of tritium.[3]

* * *

This series of elegantly brilliant experiments would stand as the pinnacle of Oliphant's hands-on research career. He later described that time of discovery that culminated in the northern spring of 1934 as 'the most thrilling' among all those in which he was directly involved at the Cavendish.

What neither he, nor those working with him, comprehended at the time was the enormity of the secret his improvised accelerator and restive inquisitiveness had uncorked.

The deuterium nuclei that Oliphant and Harteck (under Rutherford's guidance) fired at a deuterium target carried sufficiently high energy to break through electrical repulsion forces within the particles and fuse them together. This coupling of two deuterium nuclei formed the nucleus of the next-lightest element on the periodic table (helium) and released vast levels of heat and gamma radiation in the process.

The heat (in the form of particle acceleration) that was needed to overcome the electrical repulsion and achieve this result meant the process became known as 'thermonuclear fusion'. What had been successfully replicated, for the first time, at the Cavendish was the method by which the sun produces energy to sustain life on earth. It is, essentially, a giant nuclear reactor that uses hydrogen gas as fuel and huge pressure at its core to fuse around 300 million tons of hydrogen into helium each day. The energy released through this transformation creates a temperature at the sun's centre estimated to be about 20 million degrees centigrade.

It would be years before it became apparent, but Oliphant had also laid bare the secret that sits at the malignant heart of the stockpiled weapons so imminently capable of destroying the planet that the sun's radiation nurtures. While the enormous temperature and pressure generated by the sun can never be reproduced on earth, the explosion of an atomic (fission) bomb within a concentrated mass of deuterium can trigger nuclear fusion on a huge scale, which is why the hydrogen bomb (as it became known) poses such a chilling threat. In addition, the tritium that Oliphant had discovered – with Harteck's experimental input, and Rutherford's interpretive clarity – would also prove a vital ingredient in the nuclear warheads that superseded the atomic bomb.

More than once in the ensuing years, Mark Oliphant was nominated – without reward – for a Nobel Prize for his work in the successful creation, explanation and appellation of tritium. The first occasion was in 1956, when his name was put forward for the physics prize by Croatian-born Leopold Ruzicka, who had become a Nobel laureate (in chemistry) in 1939.

A second testimonial was compiled for the Nobel Foundation in 1975 by Dr John Hughes from the Australian National University's physical sciences research school, which Oliphant had founded after leaving Britain in 1950. Hughes argued that, in the wake of the first OPEC oil crisis, Oliphant's discovery of tritium raised the possibility that hydrogen might solve humans' ongoing energy dilemma: a prospect still advanced by scientists today.

> Deuterium–tritium fuel will have the lowest ignition threshold in both magnetic and inertially confined thermonuclear reactors of the future and could see mankind through a critical 50–200 year period whilst pure deuterium, and possibly hydrogen–boron, reactors are perfected.

Not only has Oliphant's discovery of tritium led to the promise of the lowest threshold fuels for thermonuclear reactors of the 21st Century but, from the viewpoint of fundamental research into the physics of super-dense matter in the laboratory, Oliphant's tritium can be compressed to a higher density than that possible with any other material under identical experimental conditions. The discovery of tritium must, therefore, be ranked as a major discovery in experimental physics which may not be fully appreciated until the 21st Century.[4]

Oliphant and Rutherford's paper detailing the meticulous processes involved in unearthing the two new isotopes and determining their respective masses has been described as being 'of little less moment than [Otto] Hahn and [Fritz] Strassmann's 1939 paper on the fission of the uranium nucleus'.[5] For the latter, Hahn would earn a Nobel Prize – albeit bestowed while he was held under house arrest in England during the final phase of the Second World War. But because Nobel Prizes are not awarded posthumously, that honour will forever elude Mark Oliphant.

After he and Rutherford finalised the paper and sent it off for publication in *Nature*, Oliphant allowed himself to bask briefly in his share of the glow of achievement that continued to illuminate the Cavendish.

'Only in the [Second World] War was I to experience such a hectic few days of work,' he would recall of the brief but brilliant period that introduced hydrogen-3 and helium-3 to the world. 'But at no other time have I felt the same sense of accomplishment, nor such comradeship, as Rutherford radiated that day [when the final draft of the paper was submitted to *Nature*].'[6]

Yet the most rewarding of Mark Oliphant's scientific triumphs had played out against the dark backdrop of desperate personal despair.

* * *

The successes Oliphant had enjoyed in the laboratory throughout physics' gilded days had brought with them further financial comfort. As a result, in late summer of 1933, he arranged for his parents to take the six-week voyage from Adelaide in order to witness first-hand the life their first-born – with his wife and son – had built in Britain.

While Rosa and Beatrice Oliphant lavished love on Geoffrey, now nearing his third birthday, Mark and Baron took the ferry from Dover for a few days' sightseeing on the Continent. It was the first weekend of autumn; the onset of the winter, from which Mark had so adamantly shielded his son during their visit to Australia two years earlier, still remained a distant enemy.

As the Oliphant men wandered among the decorated guildhalls and regal landmarks of Brussels' Grote Markt, Rosa and her mother-in-law took Geoffrey to the park near their home in the village of Grantchester (three kilometres from Cambridge, where Mark had found larger premises during his wife and son's extended stay in Australia in late 1931). The little boy ran helter-skelter across the grass, which was tinged yellow in places by the summer just passed, and beamed at his mum and grandma as he was propelled backwards and forwards on his favourite swing-set. 'He seemed as happy as a king,' Rosa would remember of that Monday afternoon, less than a week before she turned thirty.[7]

That night, Geoffrey slept in a cot that had been shifted into Rosa's room to accommodate the house guests, and she was wakened in the small hours by the chilling sounds of her son in obvious distress. The strangled grunting and rasping coming from his tiny bed were accompanied by a raging fever that Rosa detected the moment she reached down to cradle the traumatised boy.

By sunrise, his condition had worsened, and a doctor was summoned to the house from Cambridge. Rosa and Beatrice's anguish over Geoffrey's wellbeing was compounded by their bewilderment as to how they might alert Mark and Baron to the escalating crisis.

Rosa sought out her closest Cambridge friend, and Elizabeth Cockcroft rushed to be with her. Having lost their own infant son years before, the Cockcrofts inherently understood Rosa's predicament as Geoffrey continued to deteriorate. A definitive diagnosis could not be reached and, with antibiotic medicine barely grown to infancy in the 1930s, his outlook was dire.

Mark had left only a bare outline of his itinerary, and it showed plans for him and Baron to travel from Belgium's capital to Bruges on that Tuesday, 5 September. So John Cockcroft cabled an urgent message to the chief of police in the Belgian medieval market town. 'Please do utmost to find Dr Oliphant tall Australian wearing glasses arriving from Brussels today staying tonight Bruges ask to return immediately child dangerously ill. Cockcroft.'[8]

Hours dragged painfully by without word from the travellers as Rosa and Beatrice Oliphant watched Geoffrey struggle in and out of consciousness. By mid-afternoon, another telegram was sent in desperation, with a plea that the original message be broadcast on local radio, in both Flemish and English.

It met with success – though not until the following morning, when a waiter recognised the Australian physicist from the broadcast description, and alerted him to the harrowing news.

Neither Mark nor Baron had previously flown aboard a commercial airliner, but in such an emergency they raced to the nearest airfield to secure a bone-jarring flight across the Channel. Any discomfort they felt was numbed by gut-churning worry.

Upon landing in England, the distraught men jumped on the first available express train to Cambridge, and arrived to find that Geoffrey was dead. Stricken with what was formally deemed to be meningitis, he had not survived twenty-four hours from the time his symptoms first became apparent.

What had been planned as a family celebration of Mark and Rosa Oliphant's upward fortunes after those bleak early years of poverty and private traumas became, within a day, a trial of merciless grief. The suffocating sadness was exacerbated by an unspoken understanding that, in light of the health problems that had befallen Rosa in her earlier pregnancies and mindful of the tragedy visited upon Eileen Rutherford, the couple who had waited so stoically for – and then doted so dearly on – their cherished son could not consider conceiving another child.

'It was very tragic for me, but it was of course most tragic for my wife,' Mark Oliphant would recall. 'I was so upset that I hadn't been there to help her at that time …'[9]

The toll was severe on Rosa for reasons that lay deeper than the absence of her husband during those desolate hours. She had long been haunted by her own childhood memories. Her twin sister Alice's death just months after the girls were born had seen her parents' affection trained upon her older brother. When he died from typhoid aged two, Rosa became an only child who, in adulthood, regularly recalled the mantra learned from her mother and father: that her twin had been the 'pretty daughter' and Rosa was simply the one left behind.

The road from remorse lay much more clearly signposted for Mark Oliphant. It had, after all, been traversed by Cockcroft and then Rutherford in the immediately preceding years. And, in 1936, it would also be travelled by Ernest Walton upon the death of his infant son.

For each of these bereaved fathers, solace awaited behind the heavy oak gates of the Cavendish Laboratory.

12

TYRANNY'S DARK CLOUDS

Cambridge, 1933 to 1934

Following the success of the deuterium experiments, Oliphant's ambition grew to mirror his confidence. The spectacular disintegration results he had witnessed using the 'heavy' particles at comparatively low voltages led him to wonder what might reveal itself if those big bullets were fired at far greater velocities.

With Rutherford in South Africa on another of his lecture tours, and with Crowe therefore freed to act as accomplice, Oliphant constructed a high-voltage canal-ray accelerator in which deuterium ions could be absorbed completely in deuterium gas. His hope, after noting the ferocity of collisions in his earlier heavy hydrogen experiments, was that the new system would yield more energy than was applied.

These experiments indeed brought about an incandescent reaction, but not from the updated apparatus.

Rather, Oliphant's attempt to manufacture a net energy gain clashed headlong with Rutherford's publicly argued belief that the output from an atomic nucleus could not exceed what was put in.

In an address to the British Association for the Advancement of Science at Leicester in September 1933, Rutherford had famously rebutted suggestions that the atomic nucleus might one day yield limitless resources of energy for commercial and industrial use. Even if that resource could be tapped through the deployment of considerably lower voltages than were currently needed to achieve atomic disintegration, he would not be swayed from his belief that never would the nucleus's output exceed the amount of energy put in. A report in *The Times* paraphrased his outburst in Leicester that mining atoms for their riches 'was a very poor and inefficient way of producing energy, and anyone who looked for a source of power in the transformation of the atoms was talking moonshine'.[1]

Consequently, for almost the first time – though notably not the last – Oliphant inflamed his mentor's legendary ire. On returning to the Cavendish and learning of the work undertaken in his absence, Rutherford berated Oliphant for what he adjudged a flagrant waste of time. 'Surely I have explained often enough that the nucleus is a sink, not a source of energy!' he railed.[2]

As was Rutherford's way, and to Oliphant's great relief, 'he soon calmed down and we did some arithmetic. This satisfied him that I was not a fool, but foolish, and he asked me to stick to the search for facts, not fantasies.'[3]

In Oliphant's mind, Rutherford's vehemence in refuting suggestions that the energy stored within atomic structures might one day be harnessed and harvested through controlled nuclear disintegration seemed at odds with the professor's own understanding of the science – especially considering some of Rutherford's earlier calculations on the possible splitting of weightier atoms such as those of uranium. As Oliphant later noted, Rutherford had 'often speculated on the energy which could be derived from another nuclear process, the combination of hydrogen nuclei to produce heavier elements'.[4]

However, perhaps Rutherford's stubborn refusal to accept that the atom might be plundered for its still-hidden riches was not at all founded upon willing misconception. Rather, Oliphant would reflect, it may have been purposely employed due to an all too clear understanding of such a quest's consequences.

> Those who did not know Rutherford well could conclude that he seemed to be deliberately obtuse, a rare phenomenon in one whose mind absorbed so rapidly and completely any nuclear information.

I believe that he was fearful that his beloved nuclear domain was about to be invaded by infidels who wished to blow it to pieces by exploiting it commercially. Also, he disliked speculation about the practical results which could follow from any discovery, unless there were solid facts to support it.[5]

Rutherford's caution would prove well founded. The potential of those experiments that Oliphant had brazenly conducted without his director's imprimatur, and that stood no hope of succeeding at the low voltages he employed, would be exposed a decade later when the first plutonium bomb was unleashed on Nagasaki. Had Oliphant known what he was looking for, or had access to the sort of complex technology needed to find it, he might just have been that 'fool in a laboratory' against whom Rutherford would routinely warn.

Oliphant would not be shaken by Rutherford's uncharacteristic fallibility in maintaining there was nothing beyond scientific benefit to be gained from probing the atom. He would maintain an enduring assessment of Rutherford as the pioneer of the science that brought atomic energy – and with it, nuclear warfare. Later Oliphant would repeatedly take to task fellow scientists, historians and commentators who dared suggest it was Albert Einstein's relativity theory, which quantified the relationship between mass and energy, that began the race to the atomic bomb.

There were occasions when Einstein visited Cambridge and took an interest in the ongoing nuclear research work at the Cavendish. He delivered lectures to enthralled students, with Oliphant occasionally among their number.

Rutherford's distrust of theoreticians meant he would jokingly refer to Einstein's theories as 'a magnificent work of art', while laughing off any suggestions that he understood them. However, he was unashamedly inspired by Einstein's repudiation of inhumanity, and in 1933 he clearly saw the huge influence he might exert as tyranny's dark clouds gathered across Europe.

It was in that ominous year, as Germany installed Adolf Hitler as its chancellor and fell increasingly under his spell and jackboot, that Rutherford helped to found and front the Academic Assistance Council.

It was an organisation that aimed to provide support and, ultimately, safe harbour for scientists targeted by Nazi and fascist regimes. This would yield profound implications for the final outcome of the Second World War.

In May 1933, Rutherford had been approached to lend his credibility and perspicacity to the council as it aimed to provide whatever help it could, including possible repatriation. Despite his typically arduous workload, Rutherford agreed to become president of the organisation and oversee its ambitious charter to raise £1 million (almost £70 million today) to provide relief for intellectuals and teachers – mostly of Jewish heritage – rendered refugees by Hitler's race-based purges.

'Rutherford was appalled by this brutality,' recalled Oliphant, whose repugnance for Hitler's Germany also grew. 'Especially as the greatest of the German scientists, some of whom had worked with him and many of whom he knew intimately, were among the victims.'[6]

The first major fundraising initiative was an evening at London's Royal Albert Hall in October 1933, which more than 10,000 attended to hear Rutherford speak and then introduce Einstein, the event's star turn. Earlier that year, Einstein had renounced his German citizenship, and he was about to take up permanent residency in the United States. His address that evening was entitled 'Science and Civilisation', and while both he and Rutherford tactfully avoided overt mentions of politics or nationalism, they successfully rallied the crowd around the importance of ensuring that science progressed, peacefully unimpeded.

Rutherford might have held scant regard for politics and many of its practitioners, despite holding a seat in the Lords, but as Oliphant would note, he more intrinsically 'hated war and violence of every kind, though not to the point of tolerating injustice'.[7] This was the height of the great global depression, and the Cavendish's already minimalist resources under Rutherford's parsimony were stretched ever tighter. Yet those exiled scientists who sought refuge and resumption of their careers in Cambridge were welcomed without resentment or restriction by the increasingly international laboratory team.

Notable in that first wave of academic refugees was Max Born, the future Nobel laureate who in 1933 was removed from his position at

Göttingen University under Hitler's anti-Semitic race laws. Born was quickly offered a temporary position at Cambridge, which he gratefully accepted. He moved to England with his wife, Hedwig (known as Hedi), his son, Gustav, and his daughters Margarethe (known as Gritli) and Irene (who later married Cambridge-educated academic turned intelligence officer Brinley Newton-John before the couple emigrated to Australia in 1954 with their children, including daughter Olivia, who would later forge a career as a singer and actor).

However, the Borns' family dog, Trixi, was forced into quarantine by British authorities at kennels some distance from the university. Upon learning of the distress the separation had caused, the Rutherfords offered the weekend use of their car – and themselves as drivers – to convey the worried family to and from the impeached pooch. Throughout those visits, Ernest and Mary sat patiently in the vehicle, reading, until it was time for the homeward leg.

* * *

Also among the European influx that year was Austrian physicist Otto Frisch, whose grandfather was a Polish Jew. Frisch was visiting Niels Bohr at his Copenhagen Institute when he first heard rumours that the Reichstag fire in Berlin, which would be used by Hitler to invoke totalitarian rule, might have been the Nazis' own handiwork. Frisch feared the ideological madness that he had been prepared to dismiss as a fad now posed a serious and escalating threat. Soon after returning to his job at Hamburg's Institute of Physical Chemistry, Frisch quietly packed his belongings into two luggage trunks and slipped aboard a small freighter, which plied the dangerous North Sea to London.

Through his friendship with Patrick Blackett, Frisch found work at the University of London's Birkbeck College. Birkbeck was a workers' institution where Blackett had been appointed Professor of Physics earlier in the year after a falling-out with Rutherford.

It was also from Blackett that Frisch received a ticket to attend one of the Royal Institution's monthly Friday Evening Discourses. This one, in 1934, was being delivered by Rutherford, aided by his assistant Mark

Oliphant. It was an event of such prestige that men in white ties and coat-tails were delivered to the institution's brightly lit front foyer in gleaming limousines.

Clad self-consciously in the same faded brown lounge suit that he wore to the Birkbeck laboratory each day, Frisch hovered in the top row of the theatre's steeply raked seating throughout Rutherford's presentation, surreptitiously trying to secrete himself behind a pillar while still taking in the spectacle.

The demonstration he witnessed, however, would leave a profound impression. As he marvelled at the institution's famed Albemarle Street premises, Frisch beheld the passion and the pantomime that Rutherford brought to his demonstration of the Cavendish's latest atom-smashing apparatus. It was a device that Oliphant – who squirmed anxiously in his formal wear at the foot of the lecturer's platform, his back to the hushed audience – had carefully transported by car from Cambridge, and installed at the institution earlier that day.

Oliphant was fearful throughout the demonstration that the equipment he and George Crowe had so conscientiously copied from the laboratory version might misbehave in the auditorium, filled with members of the academic elite and the cream of London society. Most of them, in Oliphant's opinion, had no clue as to the intricacies or significance of elemental transmutation that Rutherford was showing.

The professor, however, was in effusive form, having earlier dined with his long-time friend Sir William Bragg, from whom he had sought out advice at Adelaide University during his maiden voyage to Cambridge in 1895. Bragg was now the Royal Institution's director, and shared Rutherford's delight at the first public demonstration of how an atom undergoes artificial transformation. It was a theatrical repeat of Oliphant's original experiment, whereby deuterium had been bombarded by deuterons to produce helium-3. As he was warmed reassuringly by the eerie light of the humming equipment, Oliphant found faint blushes of pride overriding his anxiety. He would long recall that magical evening.

> I borrowed transformers, rectifiers and some capacitors from an
> x-ray manufacturing firm in London, so that we did not need to

cart up the heavier items, and we set the whole lot up on the bench. It was touch and go whether we could get it adjusted and operating before the lecture. In case something went wrong, Crowe had a small radioactive source ready to make the counters operate at the right moment – he was adept at this kind of innocent deception.

However, all went well, and Rutherford was able to demonstrate the actual transformation of deuterons by deuterons to the audience which filled the ancient lecture room. Everyone was delighted, Rutherford most of all. The canal-ray tube glowed, the high-voltage sizzled, and the loud speaker connected with the counter and amplifier thundered with rapidly increasing speed as the voltage was raised.[8]

It was nine years, almost to the month, since Oliphant had stood similarly silent and enthralled beside a workbench at Adelaide University, as Rutherford held a far smaller audience spellbound with his stories. That was when Oliphant quietly but resolutely vowed he would one day work alongside the great scientist.

Now – on the celebrated stage of the Royal Institution's Friday Evening Discourse, treading the same worn floorboards on which Michael Faraday had unveiled to the world the power of electricity – Mark Oliphant shared the scientific limelight with Ernest Rutherford, not so much as a pupil, but as an equal.

13

THE CROWN BEGINS TO SLIP

Cambridge, North Wales and Wiltshire, 1934 to 1937

No sooner had the applause of the appreciative if largely clueless quorum at the Royal Institution faded, and the apparatus been repacked into Oliphant's bull-nosed, second-hand Morris for the return trip to Cambridge, than life there began to demonstrably change. The lustre that had come to characterise the Cavendish under Rutherford was gradually dimming – in no small part due to the once visionary director's unwillingness to grasp the future.

It wasn't only his almost paternal attachment to the atom that Rutherford jealously guarded. As nuclear physics' evolution – like Christian soldiers – marched onward, and required more powerful, more complex, more expensive machinery, Rutherford's adherence to the methods and means that had brought bygone triumphs grew ever tighter.

As a result, Cambridge's crown as the benchmark for experimental physics began to slip. And it was gleefully seized by energetic young institutions in Europe and the United States.

Particularly in the lavishly funded physics laboratories of American universities, the race to realise Rutherford's vision of particle accelerators led researchers to push the limits of experimental voltages and electromagnetic fields. Chief among them was Ernest Lawrence, at the University of California's Berkeley campus, whose development of the machine destined to redefine nuclear physics would earn him a Nobel Prize in 1939.

His initial design for a linear accelerator proved unsuccessful, but he would subsequently deliver a more far-reaching result, when he pioneered a circular model dubbed the cyclotron. This used a huge dipole magnet (essentially a giant bar magnet whose field can move particles in a circular trajectory) to bend the path of charged particles, which gained speed as they hurtled around the circular track. It was a device capable of accelerating particles to vast speeds without the need for correspondingly high voltages. As a result, Lawrence was able to dramatically increase the energy imparted to the accelerated particles, to levels that were beyond the relatively basic equipment available to Rutherford's 'boys' at the Cavendish – though Lawrence's influence on them would prove profound.

Oliphant was key among the younger minds at Cambridge who saw the future, but also understood Rutherford's reticence to yield.

> Rutherford ... disliked large and expensive equipment. He preferred to remain involved, personally, in almost all the work going on in his laboratory. His interest and ability in administration were rudimentary. He dominated the laboratory by his sheer greatness as a physicist, and provided for his colleagues and students only the very minimum of equipment required for an investigation. Rutherford, with his roots in the soil and the hard, practical life of New Zealand, bucolic in appearance, became the deep thinker and the originator of new physical concepts.[1]

* * *

Rutherford's refusal to invest in the heavy machinery needed to progress 'big science' brought other consequences too. It led, directly or otherwise, to the departure of several influential figures at the Cavendish from 1934 onward. After he had lost Blackett – who had snarled on his way out 'if physics laboratories have to be run dictatorially ... I would rather be my own dictator'[2] – another of the professor's favourite pupils, Peter Kapitza, had been refused permission by Soviet authorities to return to Cambridge after attending a 1934 conference in his native Russia.

That was around the same time as C.T.R. 'Cloud' Wilson finally retired. Before that, in 1933, Charles Wynn-Williams, whose genius for creating electronic instruments had proven crucial to Chadwick's discovery of the neutron, had taken up an assistant lecturer role in London, at the Imperial College.

The most totemic departure, however, was that of James Chadwick, Rutherford's long-time collaborator and laboratory administrator. Despite owing so much to Rutherford for the opportunities and inspiration he had provided, Chadwick eventually surrendered in the face of his professor's stubbornness. Rutherford refused to recognise that construction of a high-tension laboratory, with apparatus capable of generating up to a million volts, was essential if Cambridge were to keep pace with innovations elsewhere.

In 1935, Chadwick accepted an offer from the less renowned but more enterprising Liverpool University. He had been assured £5000 – roughly equivalent to the Cavendish's entire annual budget, and worth around £340,000 today – to build a high-energy cyclotron, as well as extra funding to maximise its impact.

'I was not prepared to quarrel with him,' Chadwick later explained as the rationale for ending his sixteen-year partnership with Rutherford and taking up the chair of physics at one of Britain's 'red-brick universities' (as the civic institutions founded in unfashionable industrial centres from the late nineteenth century became known).[3] The significance of Chadwick's decision was not lost on Oliphant, even though he would become its immediate beneficiary.

Chadwick's frustration was exemplified by a story he revealed in a letter to Oliphant thirty years after Rutherford's death. It involved a dinner that Chadwick had shared with Sir Hugh Anderson, the son of one of the founders of the Orient Company, a lucrative shipping enterprise, and who was also Master of Gonville and Caius College at Cambridge from 1912 until his death in 1928. Over a quiet meal in the college's Combination Room, Anderson told Chadwick that he greatly admired Rutherford, but was puzzled as to why the director had refused the generous grant offered him.

Nonplussed by this news, Chadwick confessed he knew nothing of the proposal, nor of Rutherford's refusal, even though he was effectively the Cavendish's chief administrator.

Anderson then went on to detail the specifics of the proposal, which he had put forward because he believed the laboratory's university funding was insufficient to allow it to conduct an optimum level of research. But he also understood the funding could not be increased, due to Cambridge's over-arching fiscal restrictions.

To help cover the shortfall, Anderson had enlisted support from a number of influential colleagues who were similarly interested in the work that Rutherford and his team were pursuing, and pulled together a purse worth £2000 a year (more than £100,000 per annum today), available within an open-ended timeframe. Rutherford only needed to advise that he required more money, and the next instalment would be forthcoming. As Chadwick recounted:

> But Rutherford did nothing, to Anderson's astonishment and indeed, exasperation. I could offer no explanation for Rutherford's refusal to say that he needed more money for research. I knew so well how hampered and restricted Rutherford was for lack of equipment and technical assistance, and his attitude was quite incomprehensible to me.
>
> 'This conversation took place some time between 1922 and 1925. It was only some years later that the explanation dawned on me ... One day I had taken Rutherford up to the [Cavendish's] Radium Room so that he could assure himself that all was in order. We had, at that time, about 400 milligrams [of] radium in solution for the preparation of radon and active deposit sources [of alpha particles].
>
> I remember well how, as we were coming down the stairs, I said that we did not have enough radium so that I had to allocate sources very carefully to meet the demands; and I said what a pity it was that somebody or other had not made a gift to him of a gram of radium ...

His reply astounded me. It was: 'well, my boy, I am very glad nobody did. Just think; at the end of every year I should have to say what I had done with it. How on earth could I justify the use of a whole gram of radium?'. This was said quite seriously.

I had, of course, long realised that Rutherford was modest about his achievements, notwithstanding his eager enjoyment of his reputation, his almost boyish delight in any laudatory references to his work, his susceptibility to flattery – aspects of his character with which you must be familiar. But I had not realised how deeply ingrained his modesty was until I pondered over this remark. And it threw a light on some arguments concerning the spending of money on research, especially on his own work.

It was then that I thought I had an explanation for his refusal to ask for the research subsidy which Sir Hugh Anderson had arranged for him; he did not feel he could justify spending so much money … on himself and his research students. And this in spite of his unique position in the scientific world, his extraordinary achievements in the past, and the urgent need to press on with the establishment of nuclear physics as a new branch of enquiry. He was, in truth, deeply modest.[4]

While humility might have partly accounted for Rutherford's aversion to private benefaction, he was also unwilling to accept direct industry sponsorship because he feared such subsidies might impugn the independence of his research work. It therefore made sense, as business and science forged a stronger nexus during the inter-war years, that Rutherford would rather go without resources than risk being beholden to commercial outcomes. In Oliphant's view, it also helped to explain why the professor was so steadfastly reluctant to consider the case for atomic energy.

'The knowledge of the inmost workings of nature, which it gave, was in many ways to him a sacred subject, not to be profaned by applications with political and industrial implications,' Oliphant later surmised. 'For him, scientific investigation was the greatest adventure of mankind.'[5]

Rutherford might not have welcomed the vested scrutiny he felt accompanied industry partnership, but the frugal ways of his regime were scarcely secret. And it was not only in the experiment rooms and on the workbenches that frayed ends were showing.

Rutherford had never been the most agile presence in a laboratory space increasingly draped in electrical wiring and housing high-voltage death traps. Now, however, the gruff clumsiness that had been cause for mutual mirth in years previous became something of an occupational hazard.

On one drizzling Cambridge morning, Rutherford lumbered up the stairs from Free School Lane to check progress among his 'boys' and hooked his damp coat over a live terminal that he mistook for a hat peg. The resulting electric shock only serving to further darken his traditionally grim morning mood.

During Oliphant's fusion experiments, Rutherford – by then in his early sixties and with hands betraying the early onset of age-related tremors – would make twice-daily inspections of the ongoing work. The first occurred late in the morning, and the second shortly before the laboratory's mandated 6pm shutdown. During those end-of-day examinations, he would impatiently seek out the gelatinous strips of photographic film used to record the charged particles expelled by exploding nuclei, as Oliphant later recounted:

> If Rutherford appeared just at the end of a run, he insisted that the record be developed as rapidly as possible, barely allowed it to be dipped in the fixing bath, and sat at the table in the next room, dripping fixing solution upon our papers and his own clothes, as he examined the tracing. His pipe dribbled ash all over the wet and sticky photographic paper. He damaged it irreparably with the stump of a pencil from his pocket, with which he attempted to mark the soft, messy material. Searching impatiently for the interesting parts of the long record, he pulled it from the coil in [George] Crowe's hands to fall to the dirty stone floor, trampling on it as he got up …
>
> We had then to do our best to finish fixing, washing and drying the paper strips, often damaged beyond repair. When it was possible, we concealed records from him till they had been properly

Ernest Rutherford in New Zealand in 1892, when he was twenty-one. Around this time he he became a boarder at the home of Mary Newton in Christchurch, and was studying for his first degree at Canterbury College. *(Cambridge Digital Library)*

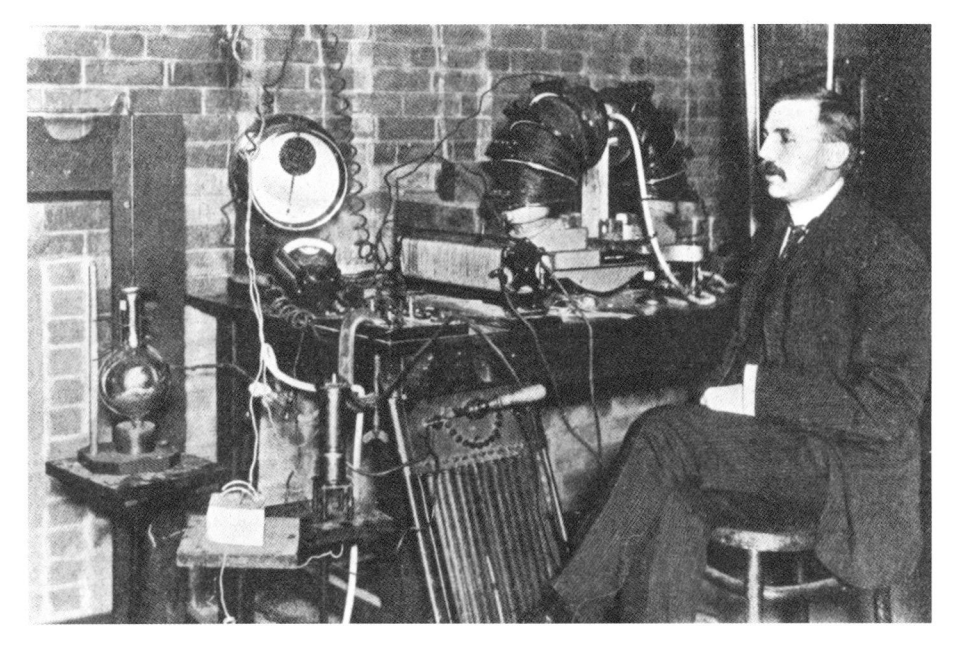

Rutherford in his laboratory at McGill University, Canada, 1905. The white cuffs beneath his jacket sleeves were borrowed from colleague Otto Hahn after the photographer said Rutherford did not appear sufficiently dignified. *(Oliphant Papers, University of Adelaide)*

Ernest Rutherford at the wheel of his first car. The 14-horsepower Wolseley Siddeley tourer was purchased with money that accompanied his Nobel Prize in 1908, and allowed his family to travel regularly between Manchester and their holiday home in North Wales. *(Oliphant Papers, University of Adelaide)*

The entrance to the Cavendish Laboratory at Cambridge, on Free School Lane. Rutherford's office was located on the second floor, while the 'garage' experimental space was in the basement and the 'nursery' training rooms were on the top level. *(Cambridge Digital Library)*

An aerial view of the Cavendish Laboratory, Cambridge, taken from the bell tower of St Bene't's Church. At left, fronting Free School Lane, is the Rayleigh Wing. The entrance gates to the laboratory are at far right. *(Cambridge Digital Library)*

The internal courtyard of the Cavendish Laboratory, as it would have appeared when Rutherford first attended Cambridge in 1895. The archway at bottom left connects to Free School Lane; the steps inside the arch lead to the building's main entrance. *(Cambridge Digital Library)*

An academic portrait of Mark Oliphant at Adelaide University, taken in the 1920s. Upon completing his honours degree in 1922, Oliphant worked as a demonstrator and lecturer in the university's Physics Department. *(Oliphant Papers, University of Adelaide)*

Harold 'Baron' Oliphant (centre) with his sons (from left) Mark, Keith, Nigel and Donald (the youngest son, Roland, was absent). The photograph was taken in 1931 after Mark returned from Cambridge University, having earned his PhD at the Cavendish Laboratory. *(Oliphant Papers, University of Adelaide)*

Ernest Rutherford at Manchester University with the apparatus he used in 1918 to fire alpha particles at nitrogen atoms. In doing so, he became the first person to split an atom, a breakthrough that heralded the birth of nuclear physics. *(Cambridge Digital Library)*

Rutherford, standing next to a seated Albert Einstein, addresses the inaugural fundraising event for the Academic Assistance Council, at London's Royal Albert Hall on 3 October 1933. The council was formed to help European scientists escape Nazi persecution. *(Oliphant Papers, University of Adelaide)*

Celyn, Rutherford's holiday cottage in the Snowdonia region of North Wales. The stone farmhouse was the site of many holidays and weekend retreats for the Rutherford and Oliphant families. *(Oliphant Papers, University of Adelaide)*

Rosa and Mark Oliphant taking the sun on the front terrace of Celyn, Mark immersed in a book while his wife combines reading and knitting. *(Oliphant Papers, University of Adelaide)*

Rutherford with Niels Bohr and (left to right) Mary Rutherford, Rosa Oliphant and Margrethe Bohr, in the garden of the Rutherfords' Cambridge home, Newnham Cottage. The photograph was probably taken by Mark Oliphant in 1930. *(Cambridge Digital Library)*

Rutherford at the beach, possibly during a family visit to England's Dorset coast in 1931. He believed that holidays should be spent away from urban environments and that a true vacation required a return to nature. *(Oliphant Papers, University of Adelaide)*

The Cavendish Laboratory team in 1932, the so-called golden year. Nine eventual Nobel Prize winners are present, eight of them in the front row, including Rutherford (centre), J.J. Thomson (with hat in hand), James Chadwick (third from left) and John Cockcroft (far right). Oliphant stands behind Thomson's right shoulder, with Ernest Walton to his left. *(Oliphant Papers, University of Adelaide)*

Rutherford at the Cavendish in 1932 with researcher Jack Ratcliffe. The illuminated sign above him warns visitors that there is sound-sensitive counting equipment nearby. It was also a not-so-subtle dig at Rutherford's thunderous manner. *(Cambridge Digital Library)*

John Cockcroft wearing headphones while counting particles at the Cavendish Laboratory. Cockcroft's work with Ernest Walton would result in the first artificial disintegration of a nucleus, which proved Einstein's famous $E = mc^2$ equation and paved the way for the atomic bomb. *(Cambridge Digital Library)*

From his 'observation tent', Ernest Walton monitors the release of fragments from atoms disintegrated by the particle accelerator he designed and built with John Cockcroft (in background, at left). The results this machine delivered earned the pair the Nobel Prize for Physics, and marked the dawn of 'big science'. *(Cambridge Digital Library)*

Rutherford with his early mentor J.J. Thomson at the Cavendish in 1933. Thomson was Cavendish director when Rutherford arrived from New Zealand in 1895, and held the position until Rutherford succeeded him in 1919. *(Cambridge Digital Library)*

Ernest Rutherford (standing) and Mark Oliphant (facing him) presenting a lecture to the Royal Institution in London, 1934. Their public demonstration of atoms splitting was the culmination of the pair's collaborative work, and a high point of Oliphant's Cavendish career. *(Cambridge Digital Library)*

Mark and Rosa Oliphant's son Geoffrey aged two, in 1933, shortly before he died. His death haunted Mark, who was overseas at the time, and forever tainted Rosa's memories of Cambridge. *(Oliphant Papers, University of Adelaide)*

Mark and Rosa Oliphant with their adopted son Michael, in North Wales in 1937. The photograph was taken during a trip to the beach around the time Oliphant ended his professional association with Rutherford and the Cavendish. *(Oliphant Papers, University of Adelaide)*

The 1-million-volt machine Oliphant sourced from Philips in the Netherlands and had assembled in Cavendish's high-tension laboratory, which opened in 1937. The 'space-age' apparatus delivered huge voltages and allowed for more powerful particle disintegration. *(Cambridge Digital Library)*

Peto, the Oliphant family home at Barnt Green in Worcestershire, England. The Oliphants had the modern extension added to the property's original gatekeeper's lodge, and Mark built the garden retaining wall and the paving. *(Oliphant Papers, University of Adelaide)*

Otto Frisch and Rudolf Peierls (left and centre), authors of the memorandum that first foresaw the possibility of an atomic bomb, with John Cockcroft (right). Frisch and Peierls fled Hitler's tyranny in Europe, and would prove vital to the Allies' Second World War victory. *(Alamy)*

Oliphant at University of California's Berkeley Radiation Laboratory in September 1941. His visit to the United States that year, and the pact he forged with Ernest Lawrence, would lead directly to the creation of the Manhattan Project. *(Oliphant Papers, University of Adelaide)*

Oliphant (left) and Ernest Lawrence in front of the 184-inch cyclotron magnet at Berkeley on 22 September 1941, during Oliphant's bomb-advocacy visit to the United States. Their friendship was decisive in the race to build an atomic bomb. *(Oliphant Papers, University of Adelaide)*

The colossal Oak Ridge complex in Tennessee, hub of the Manhattan Project's uranium-enrichment program. The facility would eventually house more than 50,000 workers, yet its purpose remained unknown to the outside world until war's end. *(Alamy)*

The Trinity Test, the first ever detonation of a nuclear bomb, in the New Mexico desert on 16 July, 1945. Head of the Manhattan Project, General Leslie Groves, described the test as 'successful beyond the most optimistic expectations of anyone'. *(Alamy)*

The shattered remains of Hiroshima, Japan, after the bombing of the city in August 1945. Oliphant was forever troubled by the impact of the weapon he helped create, and argued it should never have been deployed against civilian populations. *(Alamy)*

Mark Oliphant, John Cockcroft (centre) and J. Robert Oppenheimer at a nuclear physics conference at Birmingham University, September 1948. Cockcroft became leader of Britain's post-war atomic program, while Oppenheimer was ultimately stripped of his US security clearance. *(Oliphant Papers, University of Adelaide)*

Rosa, Michael, Mark and Vivian Oliphant aboard the *Orcades* in July 1950, leaving England for Australia. After twenty-three years abroad, Mark had been lured home to help establish the Australian National University in Canberra. *(Oliphant Papers, University of Adelaide)*

processed and measured up by us, but this was impossible when he was present while the record was being taken.

Once, at the end of a particularly heavy day, when the experiments had gone well, we decided to postpone development till next morning when we were fresh and we could handle the long strip in new developer and fixer without damage. Just as we were leaving Rutherford came in. He became extremely angry when he heard what we had decided, and insisted that we develop the film at once. 'I can't understand it', he blustered. 'Here you have exciting results and you're too damned lazy to look at them tonight'.

We did our best, but the developer was almost exhausted, and the fixing bath yellowed with use. The result was a messy record that even Rutherford could not interpret. In the end, he went off muttering to himself that he did not know why he was blessed with such a group of incompetent colleagues. After dinner that night, he telephoned me at home. 'Er, er, is that you Oliphant? I'm er, er sorry to have been so bad tempered tonight. Would you call in to see me at Newnham Cottage as you go to the Laboratory in the morning?' Next day he was even more contrite: 'Mary says I've ruined my suit. Did you manage to save the record?' He drove us mercilessly, but we loved him for it.[6]

* * *

Yet the challenges Rutherford increasingly faced were becoming emotional as well as physical. From the time of Eileen's death at Christmas 1930, he had appeared conspicuously afflicted. Arthur Eve, his former colleague from Montreal who would later comprehensively curate his personal correspondence, noted that 'the loss of his only child who he loved and admired aged Rutherford for a time; he looked older and stooped more'.[7]

The melancholy to which Rutherford would occasionally succumb following Eileen's death penetrated more deeply in 1935 when his mother Martha died in New Plymouth aged ninety-two. Although he had not seen her in almost a decade, he wrote to her every fortnight and related in intricate detail the current work in which he was immersed, and the places and people to which his lauded life had taken him.

J.J. Thomson was among the coterie of colleagues and acquaintances who sent messages of condolence to Newnham Cottage upon Martha's passing. Rutherford wrote back to the man who had effectively taken him in on his arrival from New Zealand forty years earlier: 'She was a woman of unusual strength of character [and] intelligence and all the family – particularly myself – owe much to her.'[8]

During their increasingly occasional escapes to the countryside, and at Sunday-afternoon teas in Cambridge, Oliphant saw that sadness had changed Rutherford, by then aged in his mid-sixties. 'Although normally he often dozed when sitting in an easy chair, while awake he was always busily reading or engaged in some other way. Following his mother's death he would sometimes sit staring into the distance, immobile and in deep reverie. He soon recovered, but in my view, never completely.'[9]

The holidays at Celyn came to an end in 1935. This was partly because Rutherford's incessantly hectic schedule of meetings and conferences made the lengthy road trip into the unspoiled Welsh hill country impractical. But more pragmatically, the steep trails of Nant Gwynant had become ever less appealing to Ernest and Mary, while the invasive damp and chill of the stone longhouse had finally negated its rustic charm and unimpeded views.

A search had begun the previous year for a more accessible, more comfortable alternative. After numerous reconnaissance runs into the southern counties, several acres of gently sloping pastoral land, including a grove of ancient oaks and elms, were found near the hamlet of Upper Chute. The property was on the fringe of the Wiltshire Downs, ten kilometres from Andover, in the southern corner of a larger dairy farm named, portentously, New Zealand.

Although still many hours by road from Cambridge, it was half the travel time required to reach Celyn, and Mary Rutherford enthusiastically embraced the new project. Cambridge architect Henry 'Hugh' Hughes was employed to design a simple timber bungalow that – unlike seventeenth-century Celyn – included modern plumbing and central heating, as well as upstairs lodgings for guests, and a large brick fireplace in the spacious, ground-level living room.

The Oliphants helped Mary install and arrange the furniture while Ernest was absent elsewhere, and Mark was regularly required to maintain the kitchen's paraffin stove, which could be more fractious than the lady of the house. In line with Mary's dogged belief that the remnant foundations at the foot of the property had formed part of an ancient chapel or chantry, the new residence was christened Chantry Cottage.

The Rutherfords retreated to Chantry at every opportunity, although with Ernest's schedule that became only three or four times a year. They often holidayed with the Oliphants, who also made use of the cottage when its owners could not. In recompense for that privilege, Mark was set to work in the garden whenever Mary Rutherford was in residence, and in charge.

As Oliphant later remembered, those days began 'with the physical labour of digging holes under her supervision, or lopping unwanted branches high up in trees'.

> While I was stretched dangerously at full length along a sloping branch, she would direct from the ground which canes of a rampant climbing rose she wished me to remove, not at all perturbed that I was ruining my trousers, or that it was impossible to reach one inch further with the secateurs.
>
> Although there was solid chalk beneath a very thin layer of poor soil, she insisted upon planting rhododendrons in holes which I filled with leaf mould from the woodland, but since they hate lime, they soon yellowed and died. There was a small pond among the trees where she endeavoured to establish water-loving species which I had to plant, but since it was so overhung by trees and shaded, the only result was that I was covered with mud.[10]

Rutherford himself was excused from these horticultural harangues, because they typically took place in the mornings when he was ornery. He would habitually arise in time for breakfast around 8am – bread toasted over kindling that burned low in the fireplace – and the usual reproaches from his wife that he was dribbling marmalade on his jacket, or draining his tea more rapidly than were their visitors. He would then retire to a fireside chair to browse *The Times* and the *Manchester Guardian*, or leaf through mail sent on from Cambridge. If climate or circumstances meant he was not alone in the living room, he would lead lively discussions about local history or current events.

When weather and temperament warmed, Ernest would appear in Mary's garden late in the morning at which time he was also co-opted into action. Usually he was ordered to take the end of a crosscut saw with Mark Oliphant, and set about restocking the firewood. For years, Rutherford had suffered from a partial umbilical hernia near his navel and had worn a surgical truss. It meant that the exertion of sawing was punctuated by his grunts and grimaces, as well as regular rest breaks, during which he would slump onto a log and light his pipe, using up to a dozen matches despite the tinder-dry tobacco packed into it. He would then regale Oliphant with stories of life on the farm at Fox Hill and Havelock, where, as a teenager, he found cleaving railway sleepers from native hardwood easier work than hacking at firewood in his dotage.

Yet Mark Oliphant's attachment to the Rutherfords extended further than domestic navvy and saw-pit confidant. Mary Rutherford also enlisted him on occasional covert operations, such as in early 1936, when her cousin and trusted housekeeper, Bay de Renzie, vacated her post without warning or reason. She had been in the family's employ for ten years, and Mary suspected her to be 'messing about with some man'. So she assigned the ever-obliging Oliphant to the case.

Bay had been granted use of the small car that Ernest had bought so Mary could run errands around Cambridge, and Mary became convinced the young girl was taking the vehicle to attend evening dinners and dances at hotels in surrounding villages. Her suspicions grew when she examined the household expenses and found that Bay was spending significant sums on having the car regularly tuned.

Although the closest Oliphant could claim to being a gumshoe was when the laboratory mishap meant the soles of his were vulcanised into the Cavendish floor, he did solve the mystery of the missing housekeeper. He ascertained that she was not only seeing, but had also run off with the local mechanic – who was, as suspected, also over-servicing Mary's car.[11]

* * *

Whether they were among the brutal beauty of North Wales or the more genteel Wessex landscape, the roots that linked Oliphant and Rutherford to their formative years in untamed, unhurried environments fused organically when they returned to nature. On breathless hikes across the sharp inclines of the Nant Gwynant valley, or steady ambles through the low hills of the Wiltshire Downs, it was the balm of clear air and warm sunshine, of birdcalls and soaring trees, that most readily united their kindred souls.

Rutherford's yearning to reconnect with the chalk hills and hedgerows of England's south-west grew with each administrative tussle he waged at the Cavendish. He was continuously reminded by younger researchers that, while the Cockcroft–Walton accelerator had effectively disintegrated atoms of lighter elements including lithium and boron, heavy nuclei were the next frontier.

Finally recognising the inevitable march towards big science, in 1934 Rutherford appointed Oliphant and Cockcroft to the Cavendish's building committee. The pair undertook exhaustive research to gather information on the best design and fit-out for the high-voltage facility that Cambridge would need to keep pace in the sub-atomic race. Cockcroft also began drafting letters in an attempt to secure funds from potential donors. As rival research institutions invested in more powerful particle accelerators that demanded ever-greater energy sources, members of the Cavendish team recognised the laboratory needed to match their ambition by being able to generate electricity at 1 million volts, or even 2 million, or they would risk being left behind.

The disconnect that Rutherford was beginning to feel from the science that had been his life's work was starting to wear him down. By appointing Oliphant and Cockcroft to the committee charged with investigating this next phase of the Cavendish's expansion, Rutherford had effectively outsourced due diligence on technology about which he remained volubly dubious.

The building project soaked up most of Oliphant and Cockcroft's time, and required numerous visits to institutions in Europe to inspect new installations and understand the latest thinking. It was at the Eindhoven research headquarters of Dutch technology giant Philips in early 1935 that Oliphant's eyes were opened wide to what might be possible. He was immediately taken by the potential of a 1-million-volt high-tension generator that he envisaged could be doubled in voltage capacity at the Cavendish, if housed in a suitable facility. As a commercially designed set, it would also save time and effort, as it could be shipped and installed far more easily than building such a complex piece of apparatus from scratch.

As the scope of the enterprise – the very initiative that Chadwick had been advocating until Rutherford's intransigence led him to Liverpool – grew in line with Oliphant and Cockcroft's ambitions, so too did its costs billow. Upon his return from Eindhoven, Oliphant lobbied Rutherford (who had travelled to Manchester to address the Institution of Electrical Engineers' annual dinner) with a breathless assessment of the Philips apparatus, including his estimation that a barn-like room measuring eighteen metres by twelve metres might suffice to generate 2 million volts. According to majority opinion, such a capacity was essential in order to continue the experimental work that was already foundering due to the Cavendish's inadequate infrastructure.

'It will take the best part of a year to set things up, and in the meantime we can carry on with our present apparatus,' Oliphant wrote to Rutherford, while noting that preparations for the high-voltage generator's eventual installation should begin forthwith. 'Perhaps you will feel I am over enthusiastic and inspired too much by my trip, but I have given careful thought to the question and think I have learned a great deal about it.'[12]

Within six months, however, the projected cost of this brave new addition had ballooned from £6000 to around £15,000 (approximately £1 million today), and Oliphant suddenly feared the project might be jettisoned altogether.

In August 1935 he wrote again to Rutherford, who was then on holiday in the countryside to distance himself from the tumult.

> I can see the new lab receding into the distance if we are not careful.
> Of course, if we had the money, or if you thought we could raise it,

all would be well ... I am sorry to write so fully on the matter of the new lab, but it is a thing we need urgently, and not in some distant future when all the cream has been scooped off by folks whose results we dare not trust too deeply.[13]

Then, in April 1936, Sir Herbert Austin, founder of the Austin Motor Company, wrote to Cambridge's chancellor, Prime Minister Stanley Baldwin, pledging a staggering £250,000 building endowment (almost £17 million today) to extend the Cavendish in recognition of 'the very valuable work done by Lord Rutherford and his colleagues'.[14] On face value, it seemed manna that could solve a myriad of problems besetting the laboratory. After all, as Oliphant had observed, 'Rutherford had no flair and no inclination for raising funds.'[15] He soon found that the prospect of almost inexhaustible funding would become more of a sore than a salve for the Cavendish director.

It was during a late spring weekend retreat at Chantry Cottage, shortly after Austin's offer was tabled, that Oliphant noticed his friend's mood was oddly gloomy. As the two men lolled in canvas chairs, having taken a break from Lady Rutherford's gardening duties, Oliphant pressed Rutherford for a reason. The professor conceded it was a result of 'all the goddam worry and trouble the bequest meant for him'.

> He said that he was damned if he was going to spend his time in planning a new Cavendish, that it would upset the whole lab and that Cockcroft would not do a stroke of work for a year or two.
> He went on to say that a new laboratory was unnecessary, anyway, if the number of people was kept down to about that present, and that thinking was more necessary than grand surroundings or elaborate and expensive equipment.[16]

When Oliphant dared mention that Ernest Lawrence was reporting significant results using his powerful cyclotron at Berkeley, and that the Cavendish risked being left behind by ignoring comparable technology, Rutherford snapped.

He then really blew up, becoming red in the face and shaking his pipe at me. He declared that luxury was not good for anyone, least of all a physicist, and that the amount of physics done per pound of expenditure was in inverse proportion to the total expenditure.[17]

* * *

Another matter weighing upon Rutherford during this period was the need to appoint Chadwick's replacement as his Cavendish deputy. Rutherford had also confided to colleagues, including Oliphant, that he planned to retire upon reaching age seventy, which meant a further five years at the helm. But this would require that he identify a possible successor in the interim.

In choosing his new second-in-command, he oscillated between Oliphant's irrepressible if occasionally intemperate enthusiasm, and the more staid seniority of Charles Ellis. The decision proved too difficult for Rutherford and, as a result, both men assumed the title Assistant Director of Research.

Oliphant's ordination was confirmed in a typically brief and businesslike letter from Rutherford dated 15 June 1935. A more expansive exchange took place days later, when Oliphant was summoned to Rutherford's office.

The director lit his pipe, leaned forward in a manner identical to their first meeting in that same room on a drizzling October morning eight years earlier, and imparted wisdom from which Oliphant might benefit, should he one day assume the seat on the smoking side of the desk – for which, all indicators suggested, he was now earmarked. Six years Ellis's junior, and sharing a close relationship with his director, Oliphant was a man whose star was clearly on the rise.

'You know, Oliphant,' Rutherford began, exhaling a plume of smoke that momentarily engulfed him, 'in this game it is rather important to choose the right experiments to do. But it is even more necessary to know when to stop. To understand when one should discard a line of work that has become unprofitable, for there are too many dead horses that continue to be flogged.

'Unfortunately, in industrial and governmental labs men are often assigned to a problem, or to a class of problem and have to work in that field until they retire. It's my belief that no man can make creative contributions to a subject through a particular line of attack for more than a few years. After that, he becomes stale.'

The homily hung between them, suspended in the smog, for just a moment before Rutherford forged on, stressing the need to remain open to fresh theories and approaches, using nimble and unconventional thinking. 'Don't forget,' Rutherford warned, with a vague irony lost on him in the haze, 'that many a youngster's ideas may be better than your own, and never resent the greater success of a student.'[18]

It was as if Rutherford were priming Oliphant for imminent release into a brash new world that the professor himself was afraid to confront.

Upon Mark's elevation to assistant director, the Oliphants' circumstances effectively changed overnight. They were able, at last, to move out of the rented property at Grantchester that was imbued with such desperate sadness. And they engaged Hugh Hughes – the architect who had designed Chantry Cottage – to produce a boldly modern, flat-roofed house on what had once been marshland on the city's western outskirts, a few hundred metres from where the new Cavendish Laboratory would be built forty years later.

As an homage to Mark's Adelaide Hills upbringing, the Oliphants' first property in far-away England bore the name Onkaparinga. There was little resemblance, though, between the eucalypts and baked-hard soils around Mylor, and the willow groves that flourished in the constant damp of Conduit Head Road.

Within a year, the house set on a rough cul-de-sac would echo with the sound that had been so painfully absent from Mark and Rosa's lives, when in late 1936 they finalised adoption of a four-month-old son, Michael.

The Oliphants' new, fair-haired child would never know his birth parents. But the clothes, the toys and the love that had been packed away since Geoffrey's death were heaped upon the boy whose arrival completed a family that felt truly blessed after so many bleakly trying days.

* * *

Due to his proven talent for problem-solving, and his position on the Cavendish's by now influential building committee, Oliphant began to assume a large share of the laboratory's daily responsibilities. In addition to progressing the plans for the new high-voltage facility, Oliphant was called upon to deliver lectures, supervise the work of research students, set examination papers, see to much of the day-to-day administration, and effectively oversee the experimental program.

There were times when the sheer volume of work required a touch of subterfuge. On one occasion, Oliphant was asked to pass on a technical report in order to gain Rutherford's approval: a process he saw as unnecessarily time consuming given Rutherford's similarly hectic schedule. 'Instead, Oliphant forged Rutherford's signature,' colleague Philip Dee would recall. 'And to give it verisimilitude, dribbled over the signature and rubbed it with his finger – this supposedly to represent the dripping from the end of Rutherford's pipe.'[19]

Oliphant's ascension to the post of Rutherford's most trusted deputy was confirmed in 1936, when Ellis left Cambridge to take up the physics chair at University College London.

Yet at the same time as Rutherford was grooming his young collaborator to become his successor, he effectively hastened Oliphant's departure – to his resultant fury.

* * *

As the foremost presence bestriding British physics, Rutherford was regularly sought out for recommendations to fill vacancies at rival institutions – just as Thomson had been asked to do, in recommending Rutherford for McGill in 1898.

In the early spring of 1936, with plans for the Cavendish's high-tension laboratory being furiously formulated, Rutherford was visited by Professor Neville Moss, dean of the University of Birmingham's all-encompassing faculty of science. Even after several months of searching, no suitable candidate had been found to fill the university's vacant Poynting Chair of Physics.

Moss explained that his red-brick university was shedding staff at a problematic rate, and he believed the appointment of a figure of significance from a renowned institution such as Cambridge might bolster Birmingham's prestige.

Rutherford found himself torn. The Cavendish, too, had undergone some significant personnel upheavals, and Rutherford knew his own time at the helm was ticking away. So he was initially hesitant to provide Moss with a recommendation for fear of another poaching raid, but eventually suggested he might seek out and speak with Oliphant.

Moss duly found his man and made his pitch, which included a list of reasons why Oliphant should consider quitting the bosom of nuclear research for the outer orbit of Birmingham, where facilities were even less extravagant, and there existed no formal differentiation between the disciplines of science and engineering.

Oliphant, up to his neck in plans for the new laboratory, politely heard Moss out. Then he looked around at the construction barely under way and saw reflected back an unrecognisable vision of a future Cavendish, without Rutherford's guidance. Facing a dilemma both personal and professional, he accepted Moss's offer to visit Birmingham and judge for himself the potential that had been outlined. Then he bade his visitor farewell and set off for Rutherford's office.

Oliphant was uncertain of how the professor might react to hearing of his interest in the position, but at the same time he did not seek out his mentor with trepidation. It had been Rutherford, after all, who had sent Moss his way – in much the same manner as Cox and Peterson had sought out Rutherford for the more distant posting in Montreal, when he was seven years younger than Oliphant's thirty-four. Indeed, Oliphant expected a similar response to that received by Nevill Mott – a future Cavendish professor – when he had asked Rutherford in 1932 if it would be prudent to accept the post of theoretical physics chair at Bristol University.

'Look at me,' Rutherford had enthused. 'I went to McGill and Manchester, and came back. Of course you should go!'[20]

Consequently, Rutherford's reaction left the affable Australian full of fear that their relationship, much like that of father and son, was suddenly shattered, with the force and fury of a high-speed proton on a light-metal nucleus.

'Do you mean you took seriously what Moss had to say?' Rutherford stammered, unable to conceal his sense of betrayal.

When Oliphant, taken aback, confirmed that he had, Rutherford's bluster was replaced by hurt. 'Wouldn't you rather stay here and work with me and the others?'

'Well, I'd like a show of my own. I think I'd like a show of my own,' Oliphant countered, like a naughty child pleading mitigation to escape a thrashing.

Rutherford's wounded disbelief then escalated into abject rage.

'Well, go … and be damned to you,' he roared, leaving his friend, his protégé, his assistant director of research with no option but to rise slowly from his chair and make, crestfallen, for the door.[21]

'He was very angry,' Oliphant would recall. 'Really, really nasty to me.'

But, as with so many previous outbursts, the fury soon passed, to Oliphant's great relief. 'The next day, I had a letter from him, saying he was sorry that he'd lost his temper. Wishing me all luck in what I was intending to do, and said he'd love to talk over with me … the sort of program I had in mind and so on, offering me every assistance.'[22]

The relationship had thus been quickly and fully restored, and Oliphant was left to wrestle with the offer that had come, draped in flattery, from Moss. He had not previously considered leaving Cambridge and the Cavendish. But largely due to the persuasive tone of Moss's pitch, he began to ponder the benefits of taking up a physics chair when aged in his mid-thirties, for the sake of his own scientific development as well as the financial security of his family.

In April 1936, Oliphant visited Birmingham University and took the opportunity to lay out the conditions under which he might consider accepting Moss's offer. Not only did those include a five-fold increase in the department's clearly inadequate research budget, but also a delay in taking up the appointment until the high-tension laboratory at Cambridge had been completed. That would mean any move would not occur until the latter half of 1937, allowing for a lengthy handover to further placate Rutherford.

But even when Moss agreed to those terms, Oliphant remained unsure as to the wisdom of severing his connection to the Cavendish. Ultimately, it was the calm counsel of James Chadwick, who had made a similar decision the previous year, that convinced him to take the plunge. 'Birmingham ought to have a real physics laboratory and it would be a great pity if they could not get the best man available,' Chadwick wrote from Liverpool on 17 April 1936. 'I think that means you, and most people would agree. The salary is higher than usual (and this counts in the pension too) and the conditions as far as I know are as good as anyone can expect. Such chances do not occur often.'[23]

Having reached the wrenching decision and communicated it to Rutherford, who accepted it with little outward rancour, his appointment was confirmed in June 1936, with a proposed starting date at Birmingham of October the following year. In the intervening period, there was much to celebrate during Oliphant's final phase at Cambridge. In May 1937, he learned of his election as a Fellow of the Royal Society, an honour considered second only to a Nobel Prize among British scientists. It was recognition that had been championed on his behalf by Rutherford, in his role as the revered organisation's immediate past president.

Later that year, the new high-tension laboratory was completed. With its one- and two-million-volt generators and towering roofline to encase the equipment's tall steel towers topped by bulbous spheres, the building dwarfed the Cavendish's other research spaces in both scale and investment. As Otto Frisch later recounted, the huge machines 'were well beloved by journalists and TV producers, being their idea of the shape of science to come, with their tall columns of polished metal electrodes and the crashing sparks they could be provoked to generate'.[24]

With experimental work underway using the new high-voltage equipment, Oliphant felt less anxiety, though enduring wistfulness, about his impending role at Birmingham. He was buoyed, however, by the prospect of taking charge of his own research program and facilities. And he gained reassurance from the knowledge that Rutherford remained at the Cavendish's helm, assisted by such gifted lieutenants as John Cockcroft and Philip Dee.

In October 1937, soon after he had started his tenure as Poynting Chair at Birmingham, Oliphant joined Cockcroft and Niels Bohr at a symposium in Bologna to mark the 200th anniversary of Italian physicist Luigi Galvani's birth.

It was there, in an eerie reprise of his helpless experience in Belgium when his son fell fatally ill, that Mark Oliphant's blossoming world was again plunged into gloom. A telegram sent by Philip Dee at the Cavendish alerted delegates to the unthinkable, and informed Oliphant of the unbearable. Ernest Rutherford – that force of nature, that seemingly unstoppable power, as irresistible and constant as the nuclear forces he had laid bare – was dead. At age sixty-six.

14

'REQUIEM AETERNAM'

Cambridge, 1937

Rutherford had spent September 1937 at Chantry Cottage. Oliphant was with him for some of that time, during which his defection to Birmingham was rarely raised. Rutherford was, instead, buoyant about his first visit to India, where he would preside in the new year over a joint sitting of the British Association for the Advancement of Science and the Indian Science Congress in Calcutta. He devoted much of September to preparing his welcome address, ahead of his planned departure on the lengthy sea voyage before Christmas.

While commitment to his presentation allowed Rutherford to escape the bulk of Mary's gardening demands, when he returned to Cambridge in early October, colleagues noticed his usually gregarious manner had dulled. Convinced he was merely suffering the effects of manual labour, Rutherford called upon Arthur Waller, the Victoria Avenue butcher who doubled as a backroom masseur and who, for some time, had treated the professor's troublesome knee, a painful reminder of his very first days in England.

Even for a man more trained in assessing pork bellies, it took only a cursory examination for Waller to recognise that his patient's ills required medical expertise. After swallowing castor oil, the household cure-all, Rutherford took to bed but spent the night vomiting. The family doctor was summoned next morning.

The problem was diagnosed as a strangulated hernia; a section of gut had protruded between Rutherford's abdominal muscles and caused an intestinal obstruction. It was an oft-seen condition, particularly in men who were known to undertake bouts of vigorous manual labour, such as wielding a crosscut saw, or as a result of sustained coughing fits, usually among heavy smokers.

For a typical patient, such a condition required urgent but not intricate surgery. Mary's wayward cousin and former housekeeper, Bay de Renzie, had undergone the procedure and suffered no complications. But for Baron Rutherford of Nelson, Britain's premier science peer, the case immediately became mired in the protocols of prestige.

No surgeon in Cambridge seemed willing to risk operating upon such a lauded public figure, lest the unthinkable happen and the patient be lost. Rather, the local medical fraternity opted to enlist the expertise of their more celebrated colleagues in London and further afield, several hours' train travel away.

It was the delay more than the complexity of Rutherford's condition that would prove catastrophic. For all the humility he routinely displayed, he would ultimately fall victim to his fame.

A second opinion was sought from John Ryle, Regius Professor of Medicine at Oxford University, 135 kilometres away, who then confirmed the initial diagnosis. Rutherford was admitted to Cambridge's Evelyn private hospital, but could not be operated upon until a suitably titled surgeon was summoned by Ryle from London's Harley Street.

Sir Thomas Dunhill – who, as Mary noted in updates to family, had recently treated the Princess Royal (Princess Mary) for a thyroid condition – arrived by train that Friday evening. Within two hours – but the best part of a day after the problem was identified – the Australian-born surgeon operated.

The surgery had narrowly avoided the onset of gangrene in Rutherford's gut, and Mary's early reports on his recovery were so promising that Oliphant and Cockcroft led the Cavendish delegation to Bologna in full expectation that the professor would be grumpily recuperating upon their return. But by Sunday, he began to vomit uncontrollably, and the intestinal paralysis doctors had hoped would resolve itself now appeared irreversible. A tube was inserted down Rutherford's throat to drain stomach fluids, and injections given in an effort to stimulate the inactive tract, but his condition worsened.

Come Monday, Dunhill was again dispatched to Cambridge to adjudicate on whether further surgery was required. Following a half-hour consultation, he advised Mary – who had maintained a constant bedside vigil, during which she wrote heartfelt letters of hope to her nearest, including Oliphant – that nothing more could be done. Rutherford was being kept alive by saline administered intravenously, but it was deemed too risky to proceed with another operation, as he was unlikely to survive the anaesthetic or any further disturbance to his intestine, given his fragile state.

Mary refused to abandon hope, however, and late that evening wrote to Oliphant: 'He is a wonderful patient and bears his discomforts splendidly, so tired and weary of these interminable days. There is just a thread of hope! Love to Rosie. Yours affectionately, Mary Rutherford.'[1]

The letter she began to Ernest's sister Florence (Floss) in New Zealand later that night bore greater poignancy.

> He drops out a surprisingly keen remark now and then. I could hardly bear it last night when he began to say he was sorry he was impatient sometimes, but he had really always depended on my decision of mind etc. He has enormous numbers of friends who love him, they all knew he never worked in any way for himself, always for others and he has helped all kinds of people.
>
> The doctor is here now and ... says he is downhill a little since 8am. Ernest has just told the matron that he feels a good deal better poor darling. His patience is wonderful with all these horrible tubes hanging from his mouth and his arm and he feels the heat of being well covered up and yet they dare not risk any chill.[2]

The letter remained unfinished. Further improvement never came.

In the final hours of his towering life, as well as voicing his regret that his adored wife must continue alone, Ernest Rutherford instructed her to bequeath £100 from his modest estate to Nelson College, the school that had ignited his intellect and spared him a future on a New Zealand farm.

In the early evening of Tuesday, 19 October 1937, Ernest Rutherford died with his wife at his bedside.

* * *

In Bologna, Cockcroft received Philip Dee's grim cable early the next day. The conference reconvened as Oliphant and many of his brethren made plans to return to England at the first opportunity.

Niels Bohr unsuccessfully fought tears as he told stunned delegates of overnight events. Oliphant later described Bohr's impromptu address that morning as the most unforgettably moving public tribute he would witness, and lamented that the paraphrased version published in *Nature* two weeks later captured little of its raw sincerity or heart-rending sorrow.

> With the passing of Lord Rutherford, the life of one of the greatest men who ever worked in science has come to an end. His achievements are indeed so great that, at a gathering of physicists like the one here assembled in honour of Galvani, where recent progress in our science is discussed, they provide the background of almost every word that is spoken. Rutherford passed away at the height of his activity, which is the fate his best friends would have wished for him, but … he will be missed more, perhaps, than any scientific worker has ever been missed before.[3]

Following his faltering eulogy, Bohr joined Oliphant and Cockcroft on a dash to Cambridge, their journey spent swapping treasured memories that were punctuated by lengthy reflective silences.

Back in Cambridge, pained by the guilt of once again being distantly absent when a loved one died, Oliphant insisted on bidding Rutherford a final goodbye. Accompanied by Dee, he went to the hospital morgue, 'where the body lay, pale and still … we agreed that all that made Rutherford for us had gone and only a shell remained. I was greatly distressed by this experience.'[4] Oliphant was never able to erase that upset, revisiting the sadness of his post-mortem visit thirty years later, when he confided to Dee that it had coldly clarified 'our realisation that the pallid shell was not the man we loved'.[5]

Rutherford's funeral began at noon the following Monday, a calmingly clear if slightly chill London autumn afternoon. If not for the majestic backdrop of Westminster Abbey, it might have passed as the sort of unaffected occasion that captured the man's essence. As the *Nature* eulogy recounted: 'There was no pomp or pageantry as is seen at the burial of our great naval and military leaders, no word was said of his life or achievements, but a quiet air of sincerity pervaded the whole scene and left an indelible impression that it was all as he would have wished.'[6]

Attending the coffin that contained the late physicist's ashes were Rutherford's long-time friend, Royal Society President Sir William Bragg, and his colleague and later biographer, Arthur Eve. Mary was assisted by Bohr and Oliphant throughout the short, simple service, which was attended by all from the Cavendish Laboratory, as befitting a family bereavement. The memorial was closed by a lone organist playing Basil Harwood's 'Requiem Aeternam'.

At the music's completion, Rutherford's ashes were laid to rest in the nave. Alongside the monument to Sir Isaac Newton. Near the grave of Sir Charles Darwin, and plaques honouring Lord Kelvin, James Clerk Maxwell and the late professor's inspiration, Michael Faraday.

Rutherford joined that pantheon of science immortals as the first man born in Britain's distant dominions to be granted eternal sanctuary within the abbey's holy confines.

* * *

The tributes were as profound as they were plentiful. Max Born, whose family Ernest and Mary had regularly reunited with their pet dog while it was quarantined, was a personal friend of Einstein yet rated Rutherford the greatest scientist he had known. Physicist turned novelist C.P. Snow, who had worked with Rutherford, rated him as 'very likely the major scientific figure since Newton ... It is to this man we owe the entire atomic age. And the Cavendish was the greatest physics laboratory in the world.'[7]

Oliphant believed his mentor's contributions exceeded those of Faraday and Einstein because of the influence he wielded upon a generation of students and researchers whose work, in turn, changed the world. From Frederick Soddy (1921) and Bohr (1922) through to Kapitza (1978), a dozen Rutherford disciples would earn Nobel Prizes. Six of those were awarded to Cavendish staff when Rutherford was at the helm, during which time he brought together researchers from countries as disparate as Australia and Japan, South Africa and Russia. In recognition of his breadth of achievement – from proving the earth's age to demystifying the nature of matter – he received more than 30 honorary degrees, appeared on stamps in at least four countries and eventually had an element, rutherfordium, named in his honour.

The memories that Oliphant cherished could not be quantified or catalogued. In the aftermath of the funeral, seeking solace in his new surrounds at Birmingham while quietly yearning for the comforting embrace of the not so distant Cavendish days, he would reflect on the times when Rutherford made an evening return by train to Cambridge after committee meetings or speaking engagements in London. Oliphant would collect him from the town's remote railway station, then drive him home to Newnham Cottage. It was often during these brief trips, after Rutherford had spent an hour or more collating his thoughts on the train, that the professor revealed most of himself.

On one occasion, having chaired a meeting of the Department of Scientific and Industrial Research's advisory council, Rutherford was noticeably introspective.

> When I asked whether the day had gone well, he said that he was worried because members of the Council had again attacked him for not relating the work of the Cavendish more closely to the industrial needs of the nation. Moreover, he had been accused of producing research workers who were of little or no use when faced with 'real' problems. [But] his spirits soon recovered.

> He realised, more than most, the importance of the application of scientific knowledge if Britain was to prosper, but he remained convinced that one of the best training grounds for physicists was the sort of fundamental science pursued in the Cavendish. His faith was demonstrated dramatically after his death, when the needs of [the Second World] war found almost all the men whom he had trained leading such practical developments as radar, atomic energy, and operational research.[8]

Oliphant would continue to be fiercely protective of Rutherford's reputation and influence, especially when it was suggested that Einstein laid a more credible claim to being the initial source of understanding that would ultimately deliver nuclear weapons. While acknowledging the undisputed intellect of theoretical physics' most recognisable figure, and keeping a signed letter from the German among his more prized personal papers, Oliphant took every opportunity to disavow claims that Einstein was the atomic bomb's inspiration. He used his writings 'to describe as eyewash the effort to make Einstein the father of nuclear energy. He contributed nothing. It would all have been discovered and worked out without $E = mc^2$. Einstein was one of the great men of science, but he was far from the greatest.'[9]

That honour, in Oliphant's eyes and in relation to the nuclear age's dawn, lay beyond debate. 'Much nonsense has been written and said about the early history of nuclear energy ... In fact, this achievement rests squarely upon the work of Rutherford, who discovered the nucleus, invented the methods for investigating its properties, and showed that nuclear transformations were accompanied by emission or absorption of energy enormously greater than the energies associated with chemical reactions.'[10]

Yet it was his teacher's unpretentious, unvarnished persona that resonated most deeply within Oliphant, who was similarly devoid of bombast and artifice.

> In all ways, except in his science, Rutherford was an exceptionally ordinary man in both appearance and character. Quite naturally,

he was friendly with all men and quarrelled with none. He could be moody, irritable, and on very rare occasions angry, but as with most men such deviations from the normal were rare and transient.

His success came as much from complete dedication to his work as from his innate ability, so that even the average student was inspired to emulate him. Yet, ordinary as he was, there was something in him which raised him high above others and put him in the company of the greatest men, and this something earned for him both the profound respect and deep love of all who came under his influence.[11]

Oliphant's memories were echoed by James Chadwick, Rutherford's long-time collaborator, who wrote in *The Times*:

He had, of course, a volcanic energy and an intense enthusiasm – his most obvious characteristic – and an immense capacity for work. A 'clever' man with these advantages can produce notable work, but he would not be a Rutherford. Rutherford had no cleverness – just greatness ... The world mourns the death of a great scientist, but we have lost our friend, our counsellor, our staff and our leader.[12]

* * *

Broken by Ernest's death, Mary Rutherford called upon Oliphant to sort through his papers when time permitted, while her son-in-law, Ralph Fowler, collated his correspondence. Fowler, who five years earlier had been appointed Cambridge's inaugural Plummer Professor of Theoretical Physics (the second physics chair for which Rutherford had pushed when he returned to the Cavendish in 1919), took a proprietary approach to his appointed task. Mindful that some of the private missives Rutherford had penned to Mary before their marriage had found their way into New Zealand's newspapers decades earlier, he destroyed the surviving letters he considered too personal. But among the keepsakes Oliphant retained was his own final written exchange with Rutherford.

In his formal acknowledgment of Oliphant's eventual departure from Cambridge on 8 October 1937, the professor noted his protégé's separation from the Cavendish and recorded his own gratitude for Oliphant's contribution: 'In particular, I personally appreciate how helpful you were in the whole problem of artificial transmutation and in the design of the new high-tension laboratory.' Elsewhere in the same letter, he wrote: 'You have been here at a time which has seen great developments and have taken your full part in them.'[13]

Oliphant's brief reply, rendered all the more poignant in light of events that followed, remains among his personal papers. 'You may be sure that I miss the atmosphere of the Cavendish very much,' he confessed in what would prove his parting words. 'But I feel that I may be able to disseminate a little of the enthusiasm and the method of attack in which I have been trained by you.'[14]

While Oliphant felt the loss of his mentor professionally as well as personally, his lingering hurt was not as acute as Mary Rutherford's. She responded to a note from Oliphant shortly after her husband's funeral: 'Your sweet letter is very comforting and I believe every word of it. I shall always count on you to help me. At Xmas, Rosie and you and I will carry on the good work here [at Chantry Cottage]. He had planned such a lot of jobs.'[15]

It would slowly become apparent that when Ernest died, Mary Rutherford lost not only her life's companion but also the confident, sometimes brusque personality she had developed during her thirty-seven years of marriage. Formerly so comfortable as the centre of social functions, she virtually withdrew from public life in the aftermath of her husband's death. Forced to vacate Newnham Cottage, she moved into a small flat in Cambridge, where she continued sifting through Ernest's belongings. After ascertaining that her young grandchildren were not interested, she offloaded his many degree parchments on an elderly neighbour who liked to create lampshades from historical documents. The late professor's collection of more than thirty scientific medals she donated to the University of Canterbury in Christchurch – previously Canterbury College. She returned to her former home town on New Zealand's South Island after the Second World War, where she lived alone until her death in January 1954, aged seventy-seven.

15

'A SHOW OF MY OWN'

Birmingham, 1937 to 1939

The bright new start that Mark Oliphant had sought at Birmingham was decidedly lacking in lustre. Shortly before Christmas 1937, just weeks after taking up his new post, he wrote to Niels Bohr in Copenhagen to express his lingering sadness about Rutherford's death. 'I already miss him very much as many things crop up where I require his kindly advice.'[1]

The heaviness of spirit that Oliphant felt from the passing of his friend and inspiration was not soon eased by his move. He continued to experience a deep yearning for the Cavendish, where Lawrence Bragg – the Adelaide-born son of Rutherford's close confidant, William Henry Bragg, and the youngest recipient of the physics Nobel Prize (in 1915, aged twenty-five) – would be appointed the new director in 1938. Oliphant was also dismayed by the pervading gloom of Birmingham, which had little of Cambridge's charm or academic-village feel, even though he and his family initially found rented premises in a more genteel part of the city, near the university. As Oliphant recalled years later when describing the 'drabness' of England's second-largest centre: 'There is nowhere in the city where one can stand and feel, because of the grandeur, the beauty or the character of the surroundings, that one is clearly in a great city. I know that Birmingham grew from village to city too rapidly at a time when architectural tradition was at a low ebb and when money and not beauty was the aim. But the faults do not lie entirely in the past for consciousness of the brightening effects of paint and tasteful decoration seems to exist only inside buildings and does not penetrate to the outer walls.'[2]

It was not only the freshly installed Poynting Chair of Physics who was nostalgic for life in Cambridge. Although Rosa Oliphant had initially felt few qualms about leaving behind the aching memories of her early years in England, the shift also meant vacating the recently completed Onkaparinga. So when a Birmingham University colleague of Mark's alerted the couple to the availability of a five-acre plot near the neighbouring village of Barnt Green that included a neglected main residence plus a sizeable gatekeeper's lodge ripe for remodelling, their interest was stirred.

The land was located among Worcestershire's green and gentle hills, and in what was reputedly one of the last remaining tracts of the Arden Forest. This had originally stretched as far as Stratford-upon-Avon in neighbouring Warwickshire, and local lore suggested it had inspired Shakespeare's mystical woodland fantasy *As You Like It*. Its colossal oak trees certainly exuded an ethereal aura as they rose from a summer carpet of bluebells and azaleas, or stood impassively throughout winter above clumps of aconites and snowdrops whose tones of gold and ivory shone contrastingly bright.

It was the setting, rather than the fixtures, that sealed the £2500 purchase (around £160,000 today). Unlike the serene countryside and rolling Lickey Hills around it – which the Oliphant family would explore as a panacea to the city grime twenty kilometres northward – the house replicated Birmingham University in its lack of aesthetic appeal. However, its bare brick walls, flat roof and steel-framed windows fitted with small cantilevered hoods were softened by the garden, to which the new owners became devoted, and which bloomed with beds of roses and daffodils and bluebells. It also came to include a lovingly maintained vegetable patch, and a couple of scrawny eucalyptus trees that stood as a comforting reminder of 'true home'. Further familiarity came in the name they assigned the dwelling – Peto, from the original Scottish Oliphant clan's ancient Latin motto *'Altiora peto'*, 'I seek higher things'.

In 1938, the idyllic semi-rural picture was completed when Mark and Rosa adopted a second child. Vivian Oliphant was only months old when she arrived at Peto. Born to a single mother who, as with older brother

Michael and his biological family, she would never meet, Vivian would enjoy a bountiful British upbringing, albeit at the hands of proudly Australian parents.

* * *

Now that Oliphant had settled into the physics chair in Birmingham, the realities of his new life had become formidably apparent.

Birmingham University had effectively been born in 1875 as a science training facility established by Sir Josiah Mason, whose fortune had been earned from the manufacture of pen nibs. Oliphant's first impression of his new domain was that it bore the distinct whiff of a Victorian-era relic. The entire campus, set apart from the smoke and stain of the city's heavy manufacturing heart, among the comparative gentility of Edgbaston's tree-filled estates, received its light and power via steam engines that drove a series of archaic dynamos.

The campus itself was an architectural curiosity. Its sumptuous renaissance entrance was adorned with marble columns and floors, and crowned by a soaring cupola, all within an arc of rust-red brick buildings with sandstone trim around their Tudor windows. The huge courtyard extending beyond that semicircular feature was dominated by a campanile that – at more than 100 metres high – was also the world's tallest free-standing clock tower. Modelled on Siena's Torre del Mangia, it carried the name of the university's principal benefactor and former chancellor, the Birmingham businessman turned parliamentarian Joseph Chamberlain – whose son Neville was British Prime Minister when Oliphant assumed the Poynting Chair.

Any largesse Chamberlain Senior lavished upon the institution had not found its way to the physics department, housed within the minimalist three-level structure that was the Poynting Building. Its operations also spilled over into a handful of wooden huts, initially erected as temporary measures during the First World War.

Research work was restricted to a small L-shaped laboratory notionally reserved for the Poynting Professor, a basement used primarily for storage, and a small space that was accessed through the department's cloakroom. The equally overcrowded workshop contained equipment so hopelessly out of date as to be largely useless.

When Oliphant had made his first inspection of the facilities in 1936, he had found them not only inferior to those at the Cavendish, but also funded by an annual physics budget around a quarter of what Neville Moss had suggested in the informal approach made at Cambridge. When he subsequently visited Birmingham and saw how keenly his presence was sought, Oliphant felt emboldened to lodge an ambit claim under which he might consider the move – and was then somewhat surprised when the Birmingham University council agreed.

So, in addition to his promised annual remuneration of £1300 (around £85,000 today), he was granted an initial outlay for his department of £2000, which was five times the existing yearly budget. Thereafter, he had also requested recurrent annual payments of £1000 to further help establish a nuclear physics laboratory, something glaringly absent from the Birmingham faculty in 1937. Finally, he had stipulated the employment of a full-time personal laboratory assistant to fashion and set up apparatus.

Oliphant's first order of business upon occupying the physics chair later that year was to prepare a detailed report on the state of the department for the university's council. Having endured Rutherford's reticence to seek resources for the under-funded Cavendish, which appeared lavishly appointed by comparison, he forewent subtlety in tabling his additional demands to drag Birmingham into the atomic age. His report stated bluntly: 'the facilities required for modern work in physics are completely absent. By clearing a portion of our basement and dividing it into three rooms with temporary partitions, and by clearing two rooms used for storing instruments and apparatus, temporary accommodation has been provided for some of the research but, as a result of this, I myself have nowhere to work.'[3]

He informed councillors their university could lead the world in nuclear physics – a discipline that did not even exist there in 1937 – if it outlaid £60,000 (almost £4 million today) for building and equipment upgrades.

The council unanimously acknowledged the urgent need but told Oliphant that, with Britain's economic green shoots still stunted by the great depression's shadow, it was in no position to meet his curt demand.

Instead, with bureaucratic timidity, the university announced it would be most 'grateful' to any outside entity that might fancy tipping in some cash.

In Oliphant's words: 'The Council listened with respect to what I had to say, but received my request for money in stony silence. It was clearly necessary to seek the money required for ourselves.'[4]

His initial attempts to solicit support from the factories and workshops around Birmingham, which were still struggling to regain impetus following years of economic scarcity, proved fruitless. However, after experiencing first-hand the lifeblood handed to the Cavendish to fulfil its new high-tension aspirations, Oliphant knew where to begin his tin-rattling.

Lord Austin's benevolence to Cambridge had been founded on his belief in the important work that Rutherford had been leading in the pursuit of nuclear science. Oliphant felt sure the motoring tycoon, whom he had met briefly at the Cavendish, would favourably entertain a pitch from a Rutherford acolyte looking to further that field of inquiry at the local university – particularly if that suitor was also a new near neighbour of the Austins, among the lush meadows and quiet country lanes of rural Worcestershire.

Those hopes rose higher when Lord Austin responded to an unsolicited letter with an invitation for the Oliphants to Sunday-afternoon tea. Mark and Rosa donned their formal finery and drove the four kilometres from Barnt Green to Lord Austin's Georgian manor at Marlbrook. That meeting might have failed before it began had Oliphant not realised that it might be impudent to turn into the curved gravel driveway of the Austin Motor Company founder's home at the wheel of a rattling, wheezing Morris: the brand of Austin's main marketplace rival. To avoid this potentially serious oversight, Oliphant squeezed his aged motor car against a roadside hedge not far from Austin's entrance gates, and he and his wife completed the journey to the front door on foot.

Oliphant's plan was to build rapport through sharing stories of Australia, where Lord Austin had learned engineering before he returned to England to launch his manufacturing empire. Yet his attempts at small talk over tea in the stately drawing room were cut short by Lady Austin, whom Oliphant would recall as a 'formidable woman of humble origin, with an abrupt but disarming frankness in conversation'.

'Nobody calls on us unless they want something,' she suddenly interjected. 'I suppose, like all the rest, you are after the sugar?'

Caught off-guard but red-handed, Oliphant took his chance. 'Well, eh,' he replied with a smile of admission. 'I know that your husband is interested in science and I was hoping …'

He made it no further before Lady Austin skewered him. 'We have given away all that we can afford. Would you like another cup of tea?'[5]

A final indignity came soon afterwards when, as they were being ushered to the door, Lord Austin noted the empty driveway and inquired how they intended travelling home. Determined not to reveal his source of shame lurking in undergrowth near the roadway, Oliphant fudged an answer but was immediately offered chauffeur service in His Lordship's limousine.

No sooner were he and Rosa deposited with comfort and efficiency at Peto's front door than Oliphant climbed aboard his bicycle, tucked his trouser hems into his socks and retraced the narrow laneways to the Morris. Having roped the bike to the rear luggage rack, he made dejectedly for home. With no donation in tow.

The challenges associated with 'running his own show' were by now frustratingly clear. However, Oliphant showed himself far more active than his mentor in securing resources for his work and his workers. Despite the slap-down over scones at Lord Austin's, he redoubled his efforts with the support of faculty dean Professor Moss. Through Moss's established network of contacts, the pair submitted an account of their financial plight to no less a personage than Edgbaston's parliamentary representative, and Britain's prime minister, Neville Chamberlain. Even though Oliphant felt the Conservatives' leader was 'not of a warm disposition and did not give the impression he would give much help … he did indicate that he would look into the question'.[6]

Then, within two months of Oliphant's report to the council, the university was advised by another automotive entrepreneur, Lord Nuffield, in a covering letter of 'barely five words',[7] that he would make available a grant of £60,000: the precise figure Oliphant had named. As the grateful professor wrote in a Birmingham newspaper in the days that followed, the gift represented 'an expression of his knowledge of the great role played by physical science in industry today'.[8]

The irony that the first Viscount Nuffield – born William Morris and founder of the eponymous motor vehicle manufacturer – had come to his rescue only heightened Oliphant's sense of triumph. It was a victory he felt most keenly when he climbed behind the wheel of the decrepit car he had previously considered a liability, for his daily fifteen-kilometre drive to the university. There, after seeing off bids from envious colleagues to have Lord Nuffield's gift spread across all science disciplines, the new professor began enacting his vision.

* * *

Oliphant had already announced his intention to build Britain's most powerful particle accelerator, which would shade the technology he had helped deliver to the Cavendish as well as the ongoing efforts under Chadwick at Liverpool. He also knew where a blueprint for that accelerator would be most readily found.

He duly submitted his idea to the university's newly appointed vice-chancellor, Raymond Priestley. Priestley had been the registrar at Cambridge when Oliphant first arrived there, and he was a former Antarctic exploration colleague of another Adelaidean, Sir Douglas Mawson. Oliphant received the go-ahead for a fact-finding visit to his friend Ernest Lawrence at California University's Berkeley campus near San Francisco. The sabbatical would forge a friendship that was ultimately crucial to the birth of the bomb.

Oliphant had previously met Lawrence only briefly, when the tall, almost puritanical American delivered an address at Cambridge on his way home from the 1933 Solvay Conference in Brussels. Lawrence's self-assured challenges to accepted thinking at the Solvay, a periodic gathering of the foremost international names in physics and chemistry, saw him labelled brash and impatient, qualities that had led Rutherford to observe: 'He is just like I was at his age.'[9]

From the moment that Oliphant set foot on campus at Berkeley, having sailed to America in late 1938, he was dazzled – not only by the striking similarities in character and competence between 'The Two Ernests', as he would label his memoir of working with Lawrence and

Rutherford, but also by the equally impressive research environment the former had cultivated.

Like Rutherford, Lawrence had been raised in a small, isolated community, though in South Dakota rather than the southern hemisphere. Both his parents were schoolteachers, and in adolescence Lawrence showed a mastery of mechanical equipment – motor cars, radios – and a fascination with the way the world around him worked. When he was appointed as an associate professor at Berkeley in 1928, he decided that his research focus would be the atomic nucleus. As a child of the machine age, just two months older than Oliphant, he differed fundamentally from Rutherford in his belief that answers lay in the construction of bigger, more powerful technology.

Oliphant would later write a lengthy essay comparing his two most enduring influences.

> Neither was a good speaker or lecturer, yet each influenced and inspired more colleagues and students than any other of his generation. Both built great schools of physics which became peopled with other great men, and Nobel prizes went naturally to members of their laboratories. Each was most generous in giving credit to his junior colleagues, creating thereby extraordinary loyalties.
>
> Rutherford and Lawrence were self-confident, assertive and at times over-bearing, but their stature was such that they could behave in this way with justice, and each was quick to express contrition if he was shown to be wrong.
>
> Both Rutherford and Lawrence could be devastatingly blunt and uncompromising when faced with evidence of lack of integrity, or gullibility, in scientific work. But there was one great difference. Rutherford enjoyed what has been called smoking room humour. Although his own memory for such stories was not good, his great roar of booming laughter was to be heard after dinner as he savoured the subtlety of some lewd tale.[10]

Lawrence, by contrast, never uttered a swear word and would have balked at even a suggestion of ribald humour.

His pivotal partnership with Mark Oliphant was struck over dinner in late 1938, at Lawrence's favourite restaurant, the Hawaiian-themed Trader Vic's. The pair's animated discussions then continued at the site of Lawrence's giant cyclotron at Berkeley, and in his office, where Oliphant first met the aloof Robert Oppenheimer. It was through this succession of meetings that Oliphant saw the future, in terms of both the infrastructure he would introduce to Britain, and the experimental team he would seek to assemble.

Lawrence had been drawn to exploration of the atomic nucleus as a result of Rutherford's pioneering work at the Cavendish. In turn, the American added superior engineering acumen to the late professor's canny intuition. The science behind the cyclotron – with its beams of particles harnessed by enormous magnetic forces, accelerating at increasing velocities around a curved track before being fired at targets, like stones from a slingshot – accordingly made Oliphant's eyes dance in delight.

Initially, Lawrence's machine had used an eighty-ton magnet to boost hydrogen ions, within an accelerating chamber five inches (12.7 centimetres) in diameter, to an energy level of 80,000 electron volts. More powerful magnets and more sophisticated apparatus increased the track's size to eleven inches (28 centimetres) and capability to 1 million volts. By the time Oliphant visited, Lawrence was gaining results from his thirty-seven-inch (ninety-four-centimetre) cyclotron that could accelerate alpha particles to 16 million electron volts – around four times the energy generated by alpha particles emanating from radioactive uranium. It was a precursor to the Large Hadron Collider in Switzerland, which pushes particles close to the speed of light through a subterranean ring twenty-seven kilometres long.

Oliphant sailed back to Birmingham in early 1939, as inspired by prospects of the possible as he had been after Rutherford's Adelaide visit – but this time with resources at his fingertips, and resolve born of proven achievements. This was reflected in the excited letter he wrote to Lawrence upon his return, recounting his experiences at Berkeley.

> I know of no laboratory in the world at the present time that has so fine a spirit or so grand a tradition of hard work. While there I

seemed to feel again the spirit of the old Cavendish, and to find in you those qualities of a combined camaraderie and leadership which endeared Rutherford to all who worked with him.[11]

He had decided he would use the Nuffield grant to build a sixty-inch (1.5-metre) cyclotron at Birmingham, like nothing that existed in Britain. 'Many things about the cyclotron are now clear, which were formerly hazy,' Oliphant declared elsewhere in that same letter to Lawrence. 'I return with a greater confidence and a greater belief in the cyclotron, in physics and in mankind'.[12]

That was despite the major modifications needed to the Berkeley blueprint to transplant it to Birmingham, and the unwillingness of heavy manufacturing businesses in the British Midlands to lend support through donating funds, raw materials or engineering expertise for the ambitious project. Oliphant, however, showed how relentless he could be with a mission in mind, and slowly the 'giant atom-splitting machine', as the press dubbed it, took shape.

The single-level Nuffield Building housed a pit almost five metres deep, lined with lead and surrounded by water tanks, 'so that workers in the vicinity won't be killed by the terrific radiations emitted from it', one local newspaper reported.[13] Around fifty tons of copper wire were wound into coils for the massive electromagnet, which measured almost two metres in diameter, and weighed around 300 tons. The same tonnage of plate steel was shipped from Glasgow for the magnet frame, though work stopped when a five-metre sheet fell on two research workers helping build the monster apparatus, breaking both legs of both men.

The vast copper-coil 'pancakes' that made up the magnet, so heavy they caused sections of wooden floor to give way, needed to fit together so snugly that even the smallest positioning supports proved too intrusive. Instead, the copper was lowered into place upon a layer of dry ice, which would then evaporate as the massive plates gently settled.

By May 1939, Oliphant forecast that the cyclotron would be operational by year's end. Speculation grew as to the fantastic possibilities that might emerge from its reputed power to transform every known element. 'In fact it may be practicable to change one pound of matter so that it will produce the equivalent energy of the burning of something like five million tons of coal,' a journalist enthused. 'An illustration of this possibility is that it might be possible to carry enough fuel to drive the *Queen Mary* across the Atlantic in the captain's waistcoat pocket.'[14]

* * *

The reason why Oliphant's atom-smashing machine was being touted as a source of unimaginable energy was the result of a discovery made in Germany six months earlier.

Rutherford's former McGill student Otto Hahn had been performing experiments at Berlin's Kaiser Wilhelm Institute in which uranium was bombarded with neutrons fired at slower speeds than those generated by powerful particle accelerators.

The results puzzled him, so he wrote to his friend and former collaborator Lise Meitner, an Austrian of Jewish heritage and the aunt of Otto Frisch, who had been at Rutherford and Oliphant's Royal Institution discourse years earlier. Like her nephew, who had previously fled to England by freighter in 1933, Meitner had escaped following Hitler's Anschluss of Austria in 1938 and taken up a research position at Stockholm's Nobel Physics Institute. She was similarly perplexed by the issue that had puzzled Hahn – namely, how disintegration of uranium (atomic number 92) could yield traces of barium (56), a metal so far removed from uranium on the periodic table. While transmutation was by now standard practice for nuclear physicists, it was expected that the fragments chipped off elements would reveal similar atomic properties to their source.

The exiled Meitner travelled to Kungälv, near the Swedish port city of Gothenburg, for Christmas in 1938 with family. This included her nephew, who had by now returned to Europe and the supposed safety of Niels Bohr's institute in Copenhagen. Frisch and his aunt spent the morning of Christmas Eve hiking through deep snow – Frisch wearing cross-country skis, while Meitner, who often walked more than ten kilometres per day, kept pace unaided. They would stop occasionally to rest and discuss the likely reasons for Hahn's confounding results.

It was in that postcard Yuletide setting, while they were seated upon a toppled log, that the answer dropped with the startling clarity of a fresh snow flurry.

The pair tried applying the liquid-drop model of the atom that Bohr had devised, which proposed that the nucleus of an atom resembled a water droplet held together by forces of attraction. Therefore, Bohr argued, the atom might be capable of absorbing a colliding neutral particle that carried the appropriate level of energy.

Meitner and Frisch hypothesised that Hahn's uranium atoms had done precisely that – and in so doing, had split into two roughly equivalent parts, which, because of their altered structure and the accompanying release of the particles' binding energy, instantly repelled each other with huge force.

Frisch returned to Copenhagen, where he confirmed Hahn's experimental findings. He chose the term 'fission' for this stunning process, in which one atom was transformed into two. He would later calculate that the energy released when each uranium atom split was around 200 million electron volts – enough to make a grain of sand visibly jump when a solitary uranium nucleus popped. He also noted that a single gram of uranium contained something like 2.5×10^{21} atoms – a number written as twenty-five followed by a further twenty zeroes.[15]

Frisch wrote of this find to his concert pianist mother Auguste, Meitner's sister, and conceded that when he hit upon this discovery he 'felt like someone who has caught an elephant in the jungle by the tail and did not know what to do with it. But he knew it was an elephant.'[16]

Barely a decade had passed since Ernest Rutherford stood before a meeting of the British Association for the Advancement of Science in Liverpool and dismissed suggestions that new sources of power might become available through the sudden release of energy stored in atoms. Elsewhere, Rutherford had acknowledged that, if the disintegration of heavy atoms including uranium could be somehow speeded up, so that the rate of radioactive decay could be condensed into a few days rather than millions of years, then nuclear power might be feasible. 'Unfortunately,' he had concluded 'although many experiments have been tried, there is no evidence that the rate of disintegration can be altered in the slightest degree by the most powerful laboratory agencies.'[17]

Come the end of 1938, that was palpably no longer true.

When Frisch revealed the theory to Bohr in Copenhagen, the Dane slapped a hand to his forehead before exclaiming: 'Oh what idiots we have all been.'[18] He was about to sail to the United States for a conference in Washington, DC, and when he arrived, news of the discovery roared through America's science community like a contagion.

In Berkeley, physicist Luis Alvarez was reading a newspaper in a barber's chair when he saw the headline proclaiming the splitting of the uranium atom, and bolted so quickly back to Lawrence's laboratory his barber feared his cut-throat razor had slipped. Alvarez burst in on a seminar to deliver the news to Robert Oppenheimer, who countered: 'that's impossible'.[19] Mark Oliphant first learned of the breakthrough when *Nature* magazine arrived on his desk at Birmingham in mid-January 1939.

More startling revelations had quickly followed. It was known that uranium, a metal that looks rather like lead and is softer than iron, existed as two isotopes. The more predominant of those, which had an atomic weight of 238, was found to be 140 times more abundant than uranium-235, which accounts for around 0.7 per cent of the naturally occurring element. It was Bohr who realised, while still in the United States in early 1939, that only the rare 235 isotope underwent fission when bombarded with slow neutrons.

This explained why the element remained present in the earth's primitive rocks, as well as in seawater – even though both isotopes continually emitted mild radioactivity as uranium transmuted itself to lead over billions of years. The fact that the fissionable uranium-235 existed in such minute quantities meant it was never available in sufficient concentration to promulgate violent fission and spontaneously explode.

A more sobering observation was initially made by physicist Leo Szilard, who had left his native Hungary for the United States in the aftermath of the First World War. It was at the height of the Cavendish's fame, as Chadwick discovered the neutron and Cockcroft and Walton artificially split the atom, that Szilard saw the possibility of a self-sustaining nuclear chain reaction.

If an environment could be created whereby a nucleus, split by a neutron, spat out other neutrons from the resultant disintegration, then those released neutrons could, in turn, repeat the effect as they collided. One split atom triggers another two, then four, and so on. With the right concentration of source materials to sustain that sequence, the result would be an exponential spike in atomic explosions, all taking place within millionths of a second. Each one would release vast energy in an instant, making it uncontrollable. And unthinkable.

When Szilard learned of Otto Hahn's discovery, he understood that the dystopian vision he had grimly foreseen was now at the door.

'Hahn found that uranium breaks into two parts when it absorbs a neutron,' Szilard recounted. 'When I heard this I immediately saw that these fragments, being heavier than corresponds to their charge, must emit neutrons and if enough neutrons are emitted ... then it should be, of course, possible to sustain a chain reaction.'[20] Szilard was among the first to calculate that the average number of neutrons released when a uranium atom split was 2.5.

Within a year, more than 100 scientific papers exploring uranium fission were published. Many of them touted the theoretical prospect that if sufficient quantities of the difficult to isolate uranium-235 isotope could be produced, then a bomb of cataclysmic magnitude would technically be possible. However, the consensus among those scientists postulating this macabre scenario was that such a huge volume of fissionable uranium would be needed to sustain the chain reaction – more than 100 tons, by most calculations – that a device of such size could not be delivered by any means other than a boat or train. This would rather limit its potential impact as a strike weapon. Bohr also maintained that separating such an amount of the rare uranium isotope would drain the resources of an entire nation.

Not that those likely limitations prevented newspapers and magazines from prophesying anthropogenic Armageddon. A *New York Times* headline of 29 April 1939 warned: 'Scientists Say Bit of Uranium Could Wreck New York'. A day later, across the Atlantic, Britain's *Sunday Express* trilled: 'A Whole Country Might Be Wiped Out in One Second'. This dire prognostication was followed by details of the 'first news' regarding the uranium atom split, which noted that 'if the experiments succeed, one pound of the metal will produce as much power as 20 million tons of coal do now. The new potential power was described ... by one of the scientists who are investigating it as "too great to trust humanity with".'[21]

* * *

Mark Oliphant was among those to express early fascination with the prospect of a fission bomb. Although immersed fully in the construction of his new cyclotron, Oliphant invited Niels Bohr to attend Birmingham's degree ceremony in July 1939. Bohr had recently returned from the United States and, as Oliphant spelled out in a letter to his friend, he was eager to glean the latest developments – especially amid the worrying political storm brewing in Germany.

> I am hoping that when you visit us you will bring the latest news from America. We are very interested here in the problem of nuclear fission, a question which has been taken up by the authorities here, in order that there may be no possibility missed that here is a possible source of power, or of explosion.
> I understand ... that you do not anticipate that the fission process will prove an explosive one, even under the best possible conditions, but it is felt here that the whole possibility must be investigated experimentally. As there are possibilities that the isotopes of uranium may be separated in reasonable quantity in the near future, perhaps this investigation is not so futile as may seem at first.[22]

It was the brazen aggression of Nazi Germany, whose troops had already occupied Czechoslovakia amid fears that further conquests were imminent, that underpinned the urgency to explore the prospects of an atomic weapon. But those same ever darker storm clouds gathering over Europe had already set Oliphant on an entirely different, though equally critical, scientific project.

16

THE DECISIVE DIFFERENCE

Birmingham, 1938 to 1940

As Adolf Hitler's territorial ambitions had grown, so too had Germany's military air power increased – to the point where it posed a direct danger to the island nation of Great Britain. As a consequence, from as early as 1935, British scientists and strategists had been exploring an idea that might at least provide them with a fighting chance against Germany's airborne threat.

The only public evidence of this project was the network of early warning stations – codenamed Chain Home – that were quietly installed along England's Europe-facing south coast from the middle of 1938. While the presence of sturdy timber towers reaching almost 100 metres into the sky – and connected by thick webs of electrical wiring – was obvious from miles away, their purpose stayed a tight secret.

Those privy to the significance of the system knew these towers were capable of detecting incoming aircraft up to 160 kilometres away – roughly the distance from Dover, on Kent's coast, to the French city of Lille. They were also acutely aware that, while providing up to twenty minutes' warning of incoming planes, the technology too often proved unreliable.

As war grew more likely, so too did the urgency to create a system that gave British pilots and observers more detailed information on approaching threats, and thus greater hope of repelling them.

Like so much of science, the premise that underpinned the desired outcome was infinitely simpler than the means of achieving it. The radio pulses being beamed into the skies and bounced back to earth from naturally reflective sources high in the atmosphere, and from any large objects that intervened between the two, utilised wavelengths of ten metres or more. They effectively acted as floodlights, pointed hopefully into the sky to detect any enemy aircraft that might stray into them.

If the pulses were to act more like searchlights, able to locate and lock on to imminent dangers, those wavelengths needed to be drastically shrunk. The shorter the wavelengths employed, the more information that could be collected. Under initial experimentation, the desired measure was deemed to be around a metre and a half.

If the proposed defence system were to be effective against the might of Germany's Luftwaffe, particularly if they flew missions at night, those short waves would also have to be generated with great power – by a device small enough to fit within the nose cone of a fighter plane.

* * *

The men who drew Oliphant into the closed coterie of radar – or radio direction finding (RDF), as it was known in late 1938 – held close ties to the old Cavendish network. One was John Cockcroft, the Oliphants' great friend from Cambridge. Another was Henry Tizard, scientific advisor to the Air Ministry, and chair of the secret committee charged with improving RDF's capabilities.

Tizard had been a long-time colleague and avowed admirer of Ernest Rutherford. It was through Rutherford that Oliphant had first met Tizard at Cambridge, during the Cavendish's golden period. And by late 1938, Tizard – as head of the Committee for the Scientific Survey of Air Defence – was actively seeking out universities with the capacity and the commitment to pursue short-wave research.

His approach to Oliphant proved so persuasive that, even as the new Poynting Professor was formulating his ambitious plans for a cyclotron, he was also investigating the problem posed by airborne radio transmission.

Despite the stringent confidentiality surrounding the RDF program, Oliphant's belief that the fraternal bond between scientists should transcend security restrictions saw him write immediately to the man who had replaced Rutherford as his principal confidant: Ernest Lawrence – a foreign national. As if that were not risk enough, the unclassified correspondence seeking Lawrence's input was also sent unwisely via the standard mail service.

'I have received a letter from the Defence Department in England, asking whether I will enquire about the generation of large powers of very short radio waves,' Oliphant wrote to Lawrence, even before he had returned to Birmingham from his American visit in January 1939. 'I shall be grateful, if you know this and can pass it on without betraying a confidence, if you will tell me how it is done.'[1] The letter set a pattern for Oliphant's wartime indiscretions.

Lawrence obliged with information he had gleaned from colleagues at nearby Stanford University, where the klystron – a vacuum tube used to amplify high radio frequencies – had been pioneered. Oliphant then attempted to replicate the device's capacity to generate short waves. However, given that his only previous experience with radio waves had been using an Adelaide University colleague's crystal receiver to eavesdrop on Morse code time signals and weather information, he found little success. So he returned his attention to the giant cyclotron nearing completion at Birmingham.

Tizard and Cockcroft, however, were stepping up their efforts to enlist universities' expertise as it became increasingly apparent that the policy of appeasement towards Hitler was not going to prevent another war in Europe. Britain stood a very real chance of finding itself under threat of attack. Defence of the British Isles was therefore deemed a high strategic priority. Almost 100 physicists from institutions throughout Britain were drafted and deployed in teams to study the Chain Home technology first-hand, and generate ideas on how it might be improved.

In August 1939, as one of those recruited into service, Oliphant led his unit of eight researchers to Ventnor on the Isle of Wight's southern coastline, less than 200 kilometres from France's Normandy coast.

Among Oliphant's troupe was John Randall, who had left manufacturing behemoth General Electric to pursue a Royal Society research fellowship under Birmingham's new physics professor. The group also included Harry Boot, an eager research student who confided to a newspaper at war's end that he had marvelled at the team's 'free hand to play about with a radar station'. For Oliphant, the chance to clamber across the grassed slopes of Ventnor's St Boniface Down, to scrutinise the workings of antenna towers, to comb through blueprints, and to examine hitherto unknown electrical circuitry stirred his boyhood passion for hands-on discovery.

It was while he was at Ventnor that news broke of Hitler's final act of pre-war brinkmanship. The German aggression that had escalated throughout 1939 exploded into naked hostility on 1 September, when the Nazi military invaded Poland.

Two days later, Oliphant and his team, housed in an otherwise empty hotel, listened to Neville Chamberlain's stony address to the nation on BBC Radio. For the second time in a generation, Britain was at war. Only this time, unlike the distant hostilities waged throughout Oliphant's adolescence, the conflict would be fought on his doorstep. The significance of the Prime Minister's words was not lost on the new professor.

> The Government have made plans under which it will be possible to carry on the work of the nation in the days of stress and strain that may be ahead. But these plans need your help. It is the evil things that we shall be fighting against – brute force, bad faith, injustice, oppression and persecution – and against them I am certain that the right will prevail.[2]

Oliphant knew that this would not be the boyhood adventure that the First War had seemed to him. He would need to devote all his available resources, both intellectual and physical, to the fight against tyranny. And as he gazed up at the tangle of cables that hung from the scarecrow-like timber towers, he understood that this essential first line of Britain's defence was where he could make a vital contribution.

The research at Ventnor completed, he rushed back to his Birmingham laboratory to resume the work already begun on production and detection of very short radio waves – or microwaves, as they would become more widely known.

* * *

Within a week of Poland's occupation, Birmingham's Edgbaston campus had undergone noticeable change. Barrage balloons flew above the athletics field, student accommodation was given over to medical men and nursing sisters, and a section of basement was hurriedly converted into a public safety shelter.

Blackout restrictions were enacted across Britain, and in Birmingham it was announced the university day would end at 3.30pm during winter terms to ensure that those with large distances to travel could make it home before darkness. 'If this is not done, we are likely to have more casualties from traffic accidents than from bombs,' Vice-Chancellor Raymond Priestley diarised in the war's first weeks.[3] Vehicles were now obliged to have hooded headlamps, which emitted minimal illumination. As a compensatory safety measure, white paint was applied to buffer bars and running boards, and pedestrians took to wearing luminous cards in their hatbands to lessen the chance of collisions on pitch-dark footpaths. As fears of a German land invasion rose, street signs were removed and pillboxes for machine-gunners began appearing at strategic intersections.

It was also in the early weeks of the war that Oliphant took delivery of a large manila envelope, inside which he found a second envelope stamped boldly in blood-red ink 'TOP SECRET'. It was from the Admiralty, formally requesting that Oliphant pull together a group at Birmingham to pursue development of a generator able to produce radio waves of ten centimetres in length.

Unsurprisingly, given that the prestigious wartime contract was so keenly sought by rival laboratories, including the Cavendish, Oliphant found no resistance from his vice-chancellor to the proposal that he should devote time, staff and resources to war work – even if secrecy agreements meant the university's other top brass were notionally unaware of it.

Oliphant called a meeting of almost his entire departmental staff, and set out their challenge: to unearth ways whereby vastly shorter wavelengths could be generated and utilised, not only to identify a single aircraft's approach, but for wider wartime applications such as pinpointing the direction of aerial searchlights, or improving the aim of artillery through the application of gun laying.

Although some of the groundwork had already been completed, it was abundantly clear to the entire group that the answer they sought was neither close nor simple. Since the start of the RDF research in 1935, waves had been successfully shrunk from ten to around one and a half metres, and the equipment to generate them commensurately reduced to make airborne radar a possibility – at least in theory. Yet those waves still could not be focused into concentrated beams.

To achieve that outcome, they would need to be narrowed through the use of a concave mirror, and for that apparatus to be fitted into the already cramped cockpit of a fighter aircraft and be of benefit to a pilot, the wavelength would need to be further drastically reduced – to approximately ten centimetres.

Using the same uncompromising approach he had adopted with fundraising, Oliphant launched himself into the quest to deliver microwave radar. Not only did he channel his practical problem-solving talents into this seemingly daunting project, he successfully transplanted that will and work ethic into the members of his team. His energy stemmed in part from the observation made by Charles (C.S.) Wright, another former Antarctic explorer, now the Admiralty's director of scientific research. Shortly before Germany's seizure of Poland, Wright had declared that the next war would be won by the side able to utilise the shortest radio wavelength. Wright was responsible for the development of radio valves on behalf of all Britain's armed services and, by happy chance, was also married to the sister of his one-time polar exploration colleague and Birmingham's vice chancellor, Raymond Priestley.

Oliphant assigned small groups to trial different methods of producing microwaves, and various ways of interpreting reflected signals.

His entire laboratory staff was reassigned into groups of two or three, some of which were exploring ways of improving existing technology while others were seeking altogether new ways of solving the problem that circumstances had set them.

Working with Irish physicist James Sayers, a Cambridge alumnus who was one of the few in the Birmingham laboratory with a background in radio research, Oliphant focused on adapting the klystron technology sourced from Ernest Lawrence. A second team led by Randall trialled magnetrons, high-powered vacuum tubes that used both magnetic and electric fields to generate microwaves, and that pre-dated the klystron.

The klystron model produced in the United States had yielded microwaves in the desired frequency range, but at a clearly inadequate power output of around ten watts. So while these waves could provide accurate information, they could not be used over anywhere near the distances required. Sayers worked furiously to raise that level to around 400 watts, and also found that the crystal detector – once favoured in radio sets but since regarded as obsolete – made an efficiently sensitive receiver.

As ever, despite his acknowledged absence of expertise in electrical engineering, Oliphant took a vigorous role in practical work that was not without its perils.

Britain's major cities had been provided with barrage balloons, which were attached to steel cables and flown to prevent enemy aircraft from making low-altitude bombing runs. To gauge the klystron's effectiveness in gathering waves bounced back from a solid source, its sights were to be trained on one of these balloons, which had been provided to Birmingham University at some cost, kindly borne by the institution's friends within the Admiralty.

Painstakingly filled overnight with coal gas – a precursor to natural gas in homes and workplaces – and coated with aluminium paint to replicate an aircraft fuselage, the balloon was untethered and moved to the desired testing location by Oliphant, while Sayers operated the klystron. On this breezy morning, however, Oliphant had overestimated the ground force his large frame would exert. As the dirigible caught a sudden gust, Oliphant clung to the mooring rope and was carried almost

100 metres across the university's grounds, flapping and sweating, before the wind eased and gravity returned him to earth with a thump.[4]

To further trial the klystron's capabilities, Oliphant tapped into his growing network of contacts to procure, on loan, a First World War–issue sound-locator trailer. A contraption rendered obsolete by two decades of innovation, it sported a series of oversized acoustic-amplifying trumpets, lashed to a central rotating shaft that was mounted upon a wheeled chassis. Oliphant's team stripped the relic of its original parts and fitted a pair of gleaming aluminium dishes to its central mechanism, which could be spun into an array of positions, vertically and horizontally. Its ugly functionality was captured in the name its designers bestowed upon it: 'the Dog's Breakfast'.

The portability of this crude prototype was limited by the need for continuous water flow to cool the vacuum pump, which was therefore prone to freezing in winter. Yet the Dog's Breakfast proved effective when rolled out for trials at the British naval stronghold of Portsmouth, where its microwaves proved capable of tracking even small vessels on the water, as far as the horizon.

Having served its purpose, the makeshift equipment was not greatly mourned when it was blown to pieces during one of the German air raids that would land almost 40,000 incendiary and high-explosive devices upon Portsmouth throughout the course of the war.

Years after Allied victory was achieved, Oliphant would be contacted by a pair of civil service accountants, demanding that he reconcile the outstanding loan of a 'valuable piece of government property', which he quickly established to be the Dog's Breakfast.

> They were perturbed when I said we no longer had it – that it was dead! I explained that the gigantic 'ears' and a multitude of gears and other parts [from the original First World War trailer] were in the cellar. I took them to see the heap of remnants. They wrote busily in their books, and went off without taking the parts away, apparently quite satisfied.[5]

*   *   *

While Oliphant worked with Sayers's group, Randall and his partner Harry Boot were researching magnetrons in a confined workspace tucked behind one of the physics department's lecture theatres. It was there that one of the war's most crucial developments took shape.

Randall recognised that the magnetron was currently limited by its negligible power output, barely sufficient to illuminate a hand-held torch. He also suspected that the hefty klystron would never be capable of producing sufficient power to fire out pulses that could locate an aircraft 100 or more kilometres away. He then figured that the solution might lie in combining the existing magnetron design with the properties of a resonator, an essential element of high-current devices. In the case of Randall's innovation, these resonators took the form of six cylindrical cavities drilled equidistantly around a circular block of copper, to control the frequency of the waves. It was the use of these resonators, configured to resemble the familiar cylinder that rotates bullets into the firing chamber of a revolver, that led to its name: the cavity magnetron.

At that time, late in 1939, Randall lived in a small Worcestershire village not far from Barnt Green, and would share occasional commutes to the Edgbaston campus with Oliphant in the professor's ageing Morris. During those half-hour car rides, the men discussed Randall's evolving theory on the application of the magnetron, and during one journey Oliphant agreed it was worth a trial. Randall had recently happened across a dog-eared translation of Heinrich Hertz's 1893 German text *Electric Waves* at a second-hand bookshop, a chance find that yielded inspiration's spark.

Randall figured that if the gaps between the identical cavities drilled through the copper core were carefully calibrated, they would replicate the effect of a series of side-by-side Hertz oscillators. Electrons set in motion by means of a heated tungsten-wire cathode that ran through a separate hole of the same diameter (around 1 centimetre) at the centre of the ring of cavities would be moved at high speed in a circle using a strong magnetic field. This movement caused the electrons to radiate energy, and the length of the radio waves they produced was then controlled by the size and spacing of the cavities.

Once the essential premise of the device had been proven, it took several more months of intensive practical work to ready the cavity magnetron for its all-important trial. Due to the intense heat that the magnetron produced, it was fitted with a series of copper pipes through which cold water was pumped to regulate its temperature. The entire structure also needed to be encased in copper and glass, as well as the ubiquitous sealing wax, to ensure it remained airtight when the large vacuum pumps were attached to evacuate the air from within the magnetron.

It presented a cumbersome, sprawling apparatus when it was pieced together in Birmingham's main physics laboratory, where it filled an entire workbench and sections of the surrounding floor space. But on the morning of 21 February 1940, the first cavity magnetron blazed into life as its heated discharge shimmered with a triumphant neon-blue glow. Such was the output of energy, cigarettes that jubilant research staff poked at the unit's antenna were lit almost instantaneously.

Oliphant instructed that the waves it emitted be used to illuminate a series of car headlights, to accurately gauge the power output. Then, as lamps of increasingly higher luminescence burned out and were eventually replaced by more resilient fluorescent tubes, it was assessed to be pumping out around 400 watts: ten times the output of any previous magnetron.

The other essential question – whether the wavelength would meet the Admiralty's specifications – was tested at the laboratory the following morning. The measurement of 9.8 centimetres fell comfortably within the margin of error for the ten centimetres the Admiralty had prescribed.

Subsequent engineering input from Randall's former employer, the huge General Electric Company plant at Wembley, then rendered the device more securely airtight, which meant the bulky vacuum pumps could be dispensed with. General Electric also designed sleek fins that allowed the unit to be cooled by airflow rather than constant water supply.

It was 'mission almost accomplished'. All that remained was for the unit, streamlined to be not much bigger than a modern-day, hand-held hairdryer, to undergo field testing. Oliphant himself delivered the priceless cargo, perched on the seat of his much-travelled Morris, to the radar development station secreted on the clifftop at St Alban's Head, near Swanage on Dorset's coast. From that windswept eyrie, echoes were successfully bounced off gulls flying overhead, and off a man on a bicycle several kilometres down the road. It was also successfully aimed at targets of more strategic interest, including ships in the English Channel and, vitally, a submarine periscope that had broken the waterline out at sea.

There were, of course, other obstacles to overcome before the end goal could be realised and the device installed in fighter cockpits. Further refinements were made to the cavity magnetron in the Birmingham laboratories, and in those First World War wooden huts that Oliphant had cannily appropriated to house the radar project's pre-production work. Other modifications included measures to ensure consistent energy output, as well as the manufacture of units with a range of power capacities. The biggest of them was a fifty-kilowatt version, which produced more than 100 times the energy of Randall and Boot's original model.

British pilots' dream of being able to detect and deter enemy aircraft they were unable to hear or see now appeared near to reality.

* * *

The fact that a cavity magnetron that met the Admiralty's ambitious demand had been designed and delivered within virtually six months of ceaseless work spoke volumes for Oliphant's stewardship, and the acumen and inexhaustibility of the team he had pulled together.

Yet the professor also believed that his essential role in such important work should grant him a measure of immunity from security provisions that he viewed as unwieldy and unnecessary in the pursuit of scientific outcomes. It was a view sternly at odds with that of the military brass who had commissioned the project, and for whom Oliphant was effectively working.

Throughout the First War, Ernest Rutherford had held strongly to the idealistic view that 'science is international, and long may it remain so'.[6] Oliphant wholeheartedly shared his late mentor's philosophy that the pursuit of research should benefit all, not simply those with a commercial or ideological interest

Fortified by Rutherford's stories about the different standards that applied to those in uniform and to their colleagues in laboratory coats, Oliphant had honed a healthy disregard for the military's superiority complex. He had sent ill-advised correspondence to Lawrence even before war was declared, and throughout the war years he would indignantly challenge, or often simply ignore, bureaucratic efforts to commandeer scientific discovery for partisan purposes. In his no-nonsense way, he stormed a path through what he considered the absurdities of bureaucratic red tape and overbearing officialdom.

When the British embarked on their exploration of radar in the years before the war, there had been guarded hope that the research might uncover some form of 'death ray' that could be fired from the ground to melt an aircraft's metal skin and thereby neutralise invaders in the sky.

Consequently, the harnessing of microwave radiation brought fears, which soon percolated to the public and which remain in some quarters today, that this unseen force represented the lethal radiation futuristic comic books and sensationalist magazines had warned about. To allay such fallacies when microwave generators were first manufactured in his Birmingham laboratory in 1940, Oliphant voluntarily exposed his head to a sustained bombardment of 8.3-centimetre microwaves. The demonstration continued until perspiration cascaded down his reddened face, thus proving his point that the pulses delivered nothing more sinister than heat.

His practicality also allowed him to circumvent some of the less rigorous security measures invoked during wartime. Throughout the development of the magnetron, Oliphant's primary role had been to inject ideas and remove obstacles. At one stage, his research team found need for an ingot of cadmium–copper alloy, which offered similar conductivity to pure copper, but was able to withstand more extreme temperatures. Oliphant phoned a known producer of the compound from his Birmingham office, while 'looking idly out of the window at my car, parked nearby'.

'Having confirmed that they could supply, the person at the other end asked "what is your priority number?" I had no idea that such numbers existed as we normally obtained all supplies through the Admiralty. But, without hesitation, I read off the number of my car, GOB 1676. The material was delivered promptly, and I never heard anything more about it.'[7]

As Randall would recall of his ever-present leader and enthusiastic collaborator: 'Oliphant pushed the needs of the lab without thinking whether it was the most tactful way to do it. It was just part of his personality, there was no malice.'[8]

Further evidence of his running battle with security was reported at one of the radar project's secret testing facilities on England's south coast. One of his colleagues was bailed up by a supervising officer there, who moaned: 'The guards at the gate, and the stock room people, say they can't cope. There's a man called Oliphant who keeps coming back and forth through the main gate. He doesn't have any identity card and he comes in with things marked: 'Property of the Cavendish'. He leaves [them] here, goes out, takes some of our own property for testing, as he says, yet he won't give any rhyme or reason. He just laughs it off.'

At which point Oliphant's colleague pointed out: 'Look, Sir, this is one of the great physicists of the country. His lab has produced this microwave tube which we've told you about but which you haven't yet seen. It's on three thousand megacycles …'

The superintendent's jaw dropped; his pupils dilated. 'What did you say? Three thousand megacycles?'[9]

When the science colleague confirmed that was the case, and added that the breakthrough was being heralded as the key to quelling the nightly German air raids, Oliphant was allowed to continue breezing in and out of the compound with impunity.

Yet for all its experimental success, Birmingham University's ground-breaking work brought little initial enthusiasm from the Admiralty's naval signal school, which only served to reinforce Oliphant's jaundiced perception of the military and their dismissive attitude towards science.

True to their title, the signal school preferred to communicate by hand-operated Morse keys rather than cable, and when Oliphant received a message bluntly demanding that a magnetron be tailored to a specific frequency, he politely replied that a range of models now existed, designed to fit a variety of purposes. Therefore, if the school could provide more detail as to the role they needed the device to serve, he would ensure that the appropriate version was consigned to them. He was briskly informed that the information would not be forthcoming, as the application planned was 'top-secret'. To which the professor signalled back in blunt shorthand: 'no information, no magnetron'.

Oliphant later recalled: 'Next morning the Captain of the Signal School arrived with several colleagues in uniform. They were most apologetic – arms across shoulders, much "Old Chap" and all of that. They went away with an existing magnetron, quite happy that it would do what they required.'[10]

There were occasions, however, when Oliphant was forced to concede that due process and security protocols were needed. An unannounced visit, at the height of Birmingham's magnetron development, by Lord Rothschild, who served as a senior counter-intelligence officer – and would be posthumously accused (without proof) of working as a Soviet operative – brought a sweeping inspection of the department's facilities, but no clear explanation as to the reason for his attendance.

For twelve hours or so, Oliphant believed His Lordship had failed to find any matters of immediate concern.

'Next morning, a special courier appeared with a top-secret parcel,' he later recounted. 'It contained a magnetron, which Rothschild had pocketed as he went round, and a brief note saying only "Perhaps you should tighten things up a bit!" No dressing down would have been as effective.'[11]

* * *

Albert Rowe, who would oversee Britain's radar program before being appointed Vice-Chancellor of Adelaide University, evaluated the cavity magnetron as being 'of far more importance than the atomic bomb'[12] in deciding the Second World War. It was a weapon for which Germany, despite the frightening power and technical audacity of their Luftwaffe, found no match, having based their radar system on the inferior klystron.

The cavity magnetron would go on to become a fixture on every civilian aircraft, as well as in most modern kitchens. The valve that could shoot out pulses of radio energy in ten-centimetre wavelengths and was born as a strategic weapon against imminent invasion is now the high-frequency heart of the microwave oven.

An original version of Randall and Boot's ground-breaking device is displayed in its ingenious simplicity within a glass case at London's science museum. But it was never more celebrated than when it arrived, concealed inside a securely locked and closely guarded box, in the United States at the end of summer 1940.

The German offensive that Britain had feared for more than twelve months arrived with a few sporadic sorties in early May of that year, but by August it had intensified to daily and nightly aerial raids on military installations, industrial sites and even residential neighbourhoods. Radar granted the Royal Air Force the advantage of forewarning, but the sheer scale of the onslaught meant the island's every resource was now directed towards basic survival.

With a land invasion expected at any time, it was certainly beyond Britain's compromised manufacturing capabilities to produce the cavity magnetron in the volumes that the war demanded. Therefore, barely three months after succeeding Chamberlain as Prime Minister, Winston Churchill agreed that the coveted device, so patriotically forged for the war effort, should be included among an assortment of confidential British military innovations that would be handed over, free of charge and reciprocity, to America. In return, Churchill would ask the United States – whose entry into the war remained more than a year in the future – for its essential financial and industrial aid.

This selfless gesture, or blatant cry for help, led to the Tizard Mission. It was an operation that Vannevar Bush, scientific advisor to President Franklin D. Roosevelt, would pronounce 'the most famous example of a reverse lend-lease': a reference to the program under which the United States provided aid and materiel to its allies during the war in return for local supplies and no-cost leases on military bases. It was adjudged 'reverse lend-lease' because it was one of the few items of intrinsic military value to flow into the United States in those first desperate years of the conflict.

Henry Tizard travelled to Washington, DC, on 14 August 1940, instructing his radar expert, Edward 'Taffy' Bowen, to follow by boat a fortnight later, bearing the treasure chest. In the suitcase-sized, black-lacquered deed box, whose sides had been drilled with holes to ensure it would sink without trace if it had to be hurled overboard, lay Britain's best secrets. Among them were blueprints for a new turbo jet engine, designs for a variable time fuse that enabled the programmed detonation of explosives, and plans for submarine detection devices as well as self-sealing fuel tanks.

At the bottom of the box, beneath bundles of documents and reels of film portraying the graphic horrors of Germany's relentless air blitz, rested the cavity magnetron. It would be the dramatic final presentation among Tizard's gifts of goodwill to his American science counterparts at their meeting in Washington's Wardman Park Hotel.

The effect when the magnetron was revealed and explained was even more theatrical than planned. The Americans were awestruck. The notion of airborne radar challenged their comprehension, given that their technology, like that of Germany, was built around the low-energy klystron.

James Phinney Baxter III, official historian of the Office of Scientific Research and Development, later proclaimed in his Pulitzer Prize–winning book *Scientists Against Time*:

> The British made a great advance ... but the greatest of their contributions to radar was the development of the resonant cavity magnetron, a radically new and immensely powerful device which remains the heart of every modern-day radar experiment.
>
> This revolutionary discovery, which we owe to a group of British physicists headed by Professor M.L. Oliphant of Birmingham, was the first tube capable of producing power enough to make radar feasible at wavelengths of less than 50 centimetres. When the members of the Tizard Mission brought one to America in 1940 they carried the most valuable cargo ever brought to our shores.[13]

By war's end, almost a million magnetrons had been produced, mostly in the United States and on a much smaller scale in Britain.

* * *

Even before the Tizard Mission had set out for the United States, the need to drastically bolster Britain's defensive capabilities had become critical. The eight-month 'Phoney War' that immediately followed Poland's fall, during which an anxious world observed Germany's simmering inaction with trepidation, had given way to the Nazis' brutal westward push.

Britain's vulnerability to a repeat of the aerial blitzkrieg that Hitler had unleashed on Poland meant the nation was set to high alert. Despite Churchill's defiant rhetoric, measures were being quietly taken lest its defences falter.

When news of France's capitulation was broadcast to a fearful population in June 1940, Oliphant was in his office and promptly telephoned Rosa at home in Barnt Green.

'France has fallen, you'll have to take the children and go and live in Australia,' he decreed in his authoritative tone.

'I won't go,' Rosa countered, in softly stoic defiance. She had been through a similar conversation a decade earlier, when told she was to remain in Adelaide with Geoffrey to escape Britain's winter. This time, she half-heartedly asserted that she and the children would not depart without Mark, given the political climate heading into that fraught northern summer as the German threat intensified. All the while she knew, however, that it was not a topic for debate.

Although rural Worcestershire seemed safe from the gathering terror of the Luftwaffe's strikes, Birmingham's importance as a heavy manufacturing hub meant it was already in Hitler's bombsights. With troops massed along Europe's shoreline from France to Norway, the enemy darkening the skies might also be in the streets at any time.

Within a fortnight of being told she was to leave, Rosa and the children had packed several trunks and were booked passage aboard the Orient liner *Orcades*. Mark then drove them to Southampton, where they joined the exodus of families headed, they hoped, for safer territories aboard blacked-out passenger vessels. Canada's University of Toronto had already offered to repatriate wives and children of academic staff from sister institutions in Oxford and Birmingham until it was appropriate for them to return – or so grim that their men folk would be forced to join them.

As she climbed the steep ramp slung from the cruiser, which carried around 750 passengers, Rosa clung tightly to the hands of Michael, aged four, and his two-year-old sister Vivian. Near the lower deck entrance, she turned to find her husband among the many besuited men milling fretfully dockside. 'I remember going up the gangway of the ship not knowing whether I would ever see him again,' she later conceded.[14]

By the time the *Orcades* – which two years later was sunk under attack from a German U-boat off the South African coast – had slipped moorings and slid cautiously into the Solent en route to the Channel, Mark Oliphant was on the road, heading back to Birmingham, where urgent work awaited.

The puzzle of the cavity magnetron had been largely solved, and its production effectively outsourced, but an even greater revelation had materialised in its draught.

17

'SHOULDN'T SOMEONE KNOW ABOUT THIS?'

Birmingham, 1939 to 1940

On 19 September 1939, just over two weeks after war was declared, Adolf Hitler had delivered a triumphant speech in the reclaimed Free City of Danzig (Gdansk). He taunted Britain with the clear threat that 'soon there could come a time in which we would use a weapon with which we ourselves cannot be attacked'.[1]

Some who already feared the power let loose by uranium fission, among them British scientists and politicians, interpreted this as an oblique reference to the Third Reich's atomic program, or its development of a toxic radioactive gas. Either way, the need to understand and exploit the atom's destructive power had escalated into a race for survival. Mastery of a uranium fission bomb would surely decide the war.

Physics laboratories therefore became the front line in the battle of science, and it was at Birmingham's – under the stewardship of the Poynting Professor, Mark Oliphant – that some of the most decisive blows were landed.

But while most of the physics department was consumed by the radar task, two members of staff, whom Oliphant had hand-picked during the academic exodus from Europe, were precluded from involvement once war was declared because of their status as 'enemy aliens'. Instead, they turned their agile minds to the problem that had been occupying physicists until the eruption of fighting.

* * *

Even as he was proving the theory of uranium fission, Otto Frisch was planning his escape from Copenhagen. He foresaw that the hate-driven race laws that had forced him to flee his job in Hamburg in 1933 for safety in London would soon arrive in Denmark, with the Nazis already threatening near the border. So when Mark Oliphant, on one of his sojourns in continental Europe, called in to the Bohr Institute to visit its founder early in 1939, Frisch grasped his chance. He approached Oliphant and inquired about a possible return to Britain, in whatever capacity Birmingham University might be able to use him.

Oliphant's commitment to continuing Ernest Rutherford's work was not restricted to science. He had admired the energy and empathy that Rutherford had offered the Academic Assistance Council, whose repatriation role had grown more urgent as war neared, and he met Frisch's request with unhesitating reassurance. 'Just come over,' Oliphant told the thirty-four-year-old Austrian. 'We'll find something for you to do.'[2]

Oliphant would later recall:

> The exodus of the Jews from Germany became a real river, and in Cambridge [pre-war] we received a large number of German scientists, some of whom worked with me, and I got to know them intimately and their reactions to the regime in Germany ... I learned a lot from these people, and their reactions to the treatment that they had received.[3]

However, as he had learned with his still-unfinished cyclotron, delivering on promises came with greater difficulty than proffering them. As correspondence concerning Frisch's relocation flowed between his current and prospective employers, Bohr suggested that, rather than seeking a work permit from the outset, Oliphant should offer Frisch a provisional position for several months, 'to give a few lectures ... and to assist in your work'.[4] That would necessitate only a temporary United Kingdom entry visa, which could be issued immediately. Then, if Frisch made up his mind to remain in Britain, it should prove easier for Oliphant to argue the case for permanent employment with the Home Office.

Oliphant recognised the pragmatism of Bohr's suggestion, while acknowledging that any final decision on whether Frisch might be offered a full research fellowship resided with the university's vice-chancellor. He also noted that Birmingham's physics laboratory would soon shut down for a few weeks over the 1939 summer vacation. He explained to Bohr in late May 1939:

> In writing to Frisch, I made only a very tentative offer which I thought he might be able to use to avoid the unpleasantness that might come if Denmark were invaded. I had no idea that he would feel that the sum we were able to offer would recompense him for such a change, but we would be very proud to have him here, if that would in any way help him.[5]

There was no such equivocation in Frisch's mind, however. No sooner had he received written confirmation of Oliphant's earlier informal verbal offer – and even though the details of the role he might fill at Birmingham remained unresolved – he packed two small suitcases and sailed once again for England. He arrived at the height of the Midlands summer, and spent the university break basking under the soft sun in public parks near his bed-and-breakfast lodgings.

When the students returned from holiday, he took up his notional position as auxiliary lecturer and felt an instant affection for Oliphant, whose quiet authority and cheerful energy proved to be the humanitarian balm Frisch had sought in fleeing Europe's gathering peril. He also strove to foster the same sense of egalitarian fraternity he had so cherished under Rutherford at the Cavendish.

'Oliphant had impressed me by his aura of confidence and calmness; in his presence you felt that nothing could go wrong and everything necessary would be done without fuss,' Frisch later observed.[6]

Once, as staff milled about, discussing physics during departmental morning tea, Oliphant set aside his cup and announced: 'we really must have a blackboard in here, you can't argue without one'.

Turning to a laboratory assistant, he requested that a surplus blackboard be sourced for the tea room. Upon being told all the blackboards were too large to manoeuvre up the Poynting Building's narrow, twisting stairway, Oliphant simply countered: 'never mind, we can get it through the window'.

As Frisch would remember: 'He told someone else to find a block and tackle, they went up to the roof, fixed it, hoisted the blackboard out of one window and in by the window of the tea room, and within half an hour we had a blackboard. Elsewhere that would have required days of planning and quite likely some paper work.'[7]

It was at the regular tea breaks – religiously observed by Oliphant, whose taste for sickly-sweet, currant-laced Chelsea buns had developed through the Cavendish's equivalent rituals – that Frisch first came to know another émigré staff member, Rudolf Peierls.

Born to Jewish parents in Berlin, Peierls was studying theoretical physics under Werner Heisenberg (who became his doctoral advisor) at Leipzig when he first travelled to Cambridge on summer holidays in 1928.

At first, Peierls found England a curiosity. The run-down accommodation and regimented Cambridge lifestyle, coupled with oddities including bacon-and-egg breakfasts and formal tea parties, made almost as lasting an impression as the brilliant theoretical work led by Ralph Fowler at his father-in-law's Cavendish Laboratory. Upon earning a year-long Rockefeller Fellowship in 1932, Peierls returned to Cambridge in late spring with his Russian-born wife, Genia. With Hitler's fanatical regime tightening its grip on Germany, the couple then chose to remain in Britain, where Rutherford welcomed them into the laboratory's extended family.

Through the Cavendish's tight fraternity, Rudolf Peierls became closely acquainted with Mark Oliphant, who, upon gaining the Poynting Chair, asked the German if he might be interested in joining him at Birmingham. Oliphant's informal proposal suggested that Peierls would be his new department's first academic appointment, in the role of Applied

Mathematics (essentially theoretical physics) Professor. 'The chair did not exist yet; he was trying to persuade the university to establish one and wanted to show that indeed there were suitable candidates about,' Peierls recalled.[8] Oliphant's ploy worked: the position was advertised in 1937, and Peierls was chosen as the successful applicant.

Peierls's unassuming character and utter lack of conceit led Rutherford to assert, long before the German assumed British citizenship in 1940, that he would make a perfect Englishman. (He would also be knighted decades later.) He certainly settled quickly into his professorship and, like Frisch, built a deep rapport with Oliphant, whom he later rated 'one of the great personalities of world physics'.

'He was a warm, informal and direct person with a great zest for life, a loud voice and a hearty laugh,' Peierls would say of Oliphant. 'He was happiest when he could roll up his sleeves and get to work on a piece of equipment. His instrumentation showed the influence of Rutherford's 'string and sealing wax' approach. Because of the developing needs of the experiments, Oliphant's projects were much more ambitious pieces of engineering than Rutherford would ever have contemplated.'[9]

The similarities that Peierls noted between Oliphant and Rutherford were not restricted to their practical talents. 'Oliphant was perhaps Rutherford's favourite pupil, and we understand why,' Peierls opined. 'The characters of these two men had so much in common in their directness, and in their enthusiasm for the work. Oliphant's most characteristic quality [is] the fearlessness with which he speaks his mind, no matter whether his thoughts are popular or unpopular, never calculating their effect on his image.'[10]

Both Frisch and Peierls found that the teaching duties assigned to them were not overly arduous, which led the diligent pair to seek out additional projects. Oliphant would send students from his first-year physics lectures to Frisch for expert tutelage, while Peierls used his free time to further immerse himself in the theory of nuclear reactions, founded on work published by Bohr.

As keenly as Frisch and Peierls wanted to add their expertise to the department's purportedly clandestine radar project, the repercussions for flouting restrictions that applied to their 'alien' status were potentially severe. But on many occasions when he found his will blocked by the bureaucratic obstacles of due process – as during the radar project – Oliphant would guilefully find a way around them.

Eager to engage Peierls's keen mathematical mind on problems of short-wave radio pulses, Oliphant would approach the German during tea breaks and pose apparently rhetorical questions that both men understood related directly to the highly confidential research.

'If you were faced with the problem of solving Maxwell's equations for a cavity with conducting walls in the shape of a hemisphere,' Oliphant would muse, distractedly stirring his tea 'could you cope with it?'

To which Peierls, playing his pantomime role as if the query were no more than social small talk, would respond: 'Well, it's an interesting problem; I'll give it some thought.'

A few days later, as they gathered over more Chelsea buns, Peierls would suddenly recall the earlier conversation and announce: 'I have a solution to that problem you gave me …'

As Frisch later confirmed:

Peierls knew that this was connected with the generation of very short electrical waves, such as were needed for radar, and Oliphant knew that Peierls knew, and I think Peierls knew that Oliphant knew that he knew. But neither of them let on; they both pretended that this was purely an academic problem that had occurred to Oliphant out of the blue …[11]

To keep Frisch and Peierls more productively occupied, Oliphant felt that uranium fission would prove a suitably benign alternative research focus, given that subject was not thought likely to deliver any immediate impact on the war now under way.

Frisch the experimentalist therefore resumed the work he had abandoned upon fleeing Denmark.

* * *

Frisch began by testing Bohr's proposition that it was not the common uranium-238, but the rare 235 isotope that fissioned when bombarded by slow neutrons. It was work that had previously been started at Birmingham by Philip Moon, whom Rutherford had dubbed 'Oliphant's Satellite' during their research days at Cambridge, but who had found little success in his uranium endeavours.

In order to try to separate a usable quantity of the 235 isotope, which comprised such a tiny proportion of naturally occurring uranium, Frisch required a greater supply of the metal than Britain's air ministry had provided for Moon's earlier research.

Oliphant, knowing who would be best placed to authorise the release of this highly prized and closely guarded material, wrote to Henry Tizard late in 1939 requesting five hundredweights (around 250 kilograms) of uranium oxide.

Tizard received Oliphant's message shortly after entertaining a similar request from James Chadwick at Liverpool University, where experiments on uranium fission had begun within days of the war's commencement. Knowing that Oliphant's laboratory was fully engaged in radar research, Tizard inquired as to who at Birmingham would be leading investigations in the sensitive field of nuclear fission.

'I am told you have refugees in your laboratory,' Tizard wrote, 'and it just occurs to me that perhaps it is a little hard on the English physicists who are interested in the same problem, but who are now deeply engaged on war work, if the refugees get a good start on them by being in the fortunate position of being able to devote their time to pure science.'

Oliphant's retort pulled no punches. He advised Tizard that the work would be undertaken by Otto Frisch who, a year earlier, had worked out the theory of fission in collaboration with his celebrated aunt, Lise Meitner. Then, for good measure, he added: 'In my opinion, it is much more important that work of this nature should be done than that any question should be raised about whose effort is employed to get the answer.'[12]

The uranium oxide was duly delivered, and although the Birmingham laboratory's already strained resources were devoted foremost to the radar program, Frisch set about making plans to separate the elusive 235 isotope using equipment devised by German physicist Klaus Clusius.

The Clusius apparatus was a long, vertical glass tube, through which ran a metal wire that could be heated electrically. The tube would then be filled with a gaseous compound of the element to be separated, with the lighter 235 isotopes gathering at the top (where it was hotter) while the heavier ones collected around the cooler base. The immediate difficulty in that process would be procuring the laboratory's sole glass blower, who fielded almost constant demands from the radar team.

To ensure that Frisch was not rendered idle while waiting for the apparatus, Oliphant – who had already secured an unused lecture room in which Frisch could conduct his practical work – handed him another task: to prepare an article encapsulating the current state of nuclear physics for the Chemical Society's 1939 annual report.

As Frisch would remember:

> I managed to write that article in my bed-sitter where in daytime, with the gas fire going all day, the temperature rose to 42° Fahrenheit (about 6° Centigrade) while at night the water froze in the tumbler by my bedside. What I did was to pull a club chair up close to the gas fire, wear my winter coat and put the typewriter in my lap so as to be protected from all sides; the radiation from the gas fire stimulated the blood supply to my brain …[13]

In his report, Frisch revisited Bohr's belief that while a chain reaction of uranium isotopes split by neutrons – with the resultant release of multiple neutrons able to promulgate further fission – was distinctly possible, the likelihood of a violent explosion appeared fanciful. The secondary neutrons released would move too slowly to replicate the cleaving process in the predominant uranium-238 isotope, and only a small fraction of the available uranium-235 atoms would thereby undergo fission.

It had already been shown that slowing the release of neutrons through the use of a moderating agent (such as heavy water or graphite) increased the likelihood of achieving fission in uranium-235. However, that process would also quell the speed at which any self-sustaining fission chain reaction might take place. This suggested nuclear fission was potentially useful as a constant energy source, but if employed in a bomb, fission would likely only replicate the sort of energy burst gained from igniting a similarly large stockpile of gunpowder.

Having submitted his article, which essentially dismissed the notion of an imminent uranium bomb, Frisch returned to isotope separation. Yet his interest was now piqued by the question of how much uranium-235 isotope might be required to sustain such a chain reaction. Given that uranium-235 and uranium-238 shared the same properties and were only distinguishable by mass, they could not be separated by simple chemical methods; that outcome would have to be achieved by physically splitting the lighter (235) isotope from the heavier (238) particles.

This would require complex mathematics – and the handily available input of Peierls.

* * *

From the time Oliphant had united them in the sanctuary of Birmingham, Frisch and Peierls had been drawn closer together by their deep resentment of Nazism. But the bond they shared transcended their cultural pedigrees and recent painful experiences. Frisch would take every opportunity to escape his frigid lodgings and spend evenings with Peierls and his wife, and when the couple moved to larger premises elsewhere in Edgbaston, Frisch gratefully accepted their offer of permanent accommodation.

The trio would routinely embark on long, exploratory walks through the Midlands, occasionally stopping overnight at village hotels and resuming their trek next morning.

It was on one of those outings, in May 1940, that the three émigrés walked into a pub on a spring bank holiday Monday to find the clientele within all huddled in grim-faced silence. They were listening intently to the landlord's radio set, from which the voice of freshly installed prime minister Winston Churchill informed his anxious nation that he had nothing to offer them 'but blood, toil, tears and sweat'.[14] Even though the trio of travellers had been refused overnight accommodation in the village, and the local constable had declined their suggestion they sleep in the otherwise unoccupied police cells, they stayed for the duration of the famous address. 'Nobody left until Churchill had finished,' Peierls remembered.[15]

Earlier in the war, Peierls had been approached by British intelligence personnel for his opinion on how the activities of Germany's leading physicists might best be monitored. It was an attempt to ascertain how far the Third Reich had progressed towards harnessing the frightening power within uranium atoms.

Peierls had suggested keeping track of their published research output, which would verify whether they were still in their usual places of work or had been collectively sequestered elsewhere. He then offered to study the German-language science journals as they became available, for that purpose. It was the worryingly real fear that Hitler was engaged, and possibly ahead, in the race for the bomb that drove the pair relentlessly on.

While Frisch was focused on isotope separation, Peierls – who had so craved a part in Britain's anti-war effort that he had successfully applied to join Birmingham's auxiliary fire service – immersed himself in fission theory. He understood that if the mass of material was too small, there was a high chance that the neutrons would escape without initiating secondary fission, thus rendering the material inert. Too large a mass, and the process would race away uncontrolled and effectively self-destruct.

Producing a sustainable, managed chain reaction would require a very specific volume of fissionable material. The thesis on which Peierls based his initial calculations, published by French theoretical physicist Francis Perrin earlier in 1939, estimated that amount to be forty-four tons. If, however, the uranium was encased within lead or iron, from which neutrons that escaped might be reflected back into the reaction, it could possibly be reduced to thirteen tons.

Perrin's calculations were loosely supported by James Chadwick at Liverpool, who had advised Tizard that it could be anywhere between one and forty tons. This reinforced the idea put forward by Niels Bohr, and reiterated by Frisch in his report for the Chemical Society, that the size of any weapon utilising a uranium chain reaction would be so huge that it could not be transported by aircraft, and was therefore of little practical use.

However, by improving the calculations in Perrin's paper, Peierls showed that the mass might be significantly lower, though still in the order of tons. He discussed his findings with Frisch, who saw no issue with making them public, given that they echoed the existing literature, which ruled out the likelihood of a bomb. Peierls therefore submitted his paper for publication by the Cambridge Philosophical Society.

Both men were now on the public record as having dismissed the prospect that uranium fission might fuel a bomb, but their search continued. Frisch focused on the one unexplored method by which an explosive chain reaction might be set off in uranium. It had been ruled out as unachievable through the firing of fast and slow neutrons at uranium-238. And his own recently published report showed that *slow* neutron bombardment of uranium-235 might yield energy for power generation, but at an insufficient rate for use as a weapon. The only outstanding scenario was to consider the effect that *fast* neutrons unleashed upon uranium-235 might bring.

'I wondered – assuming that my Clusius separation tube worked well – if one could use a number of such tubes to produce enough uranium-235 to make a truly explosive chain reaction possible, not dependent on slow neutrons,' Frisch later recalled. 'How much of the isotope would be needed?'[16]

The scarcity of insight into uranium-235 meant that Frisch had to estimate a crucial detail, the fission cross-section of the difficult to obtain isotope, so that he could insert it into the mathematical formula that Peierls had refined in his paper. A cross-section is a measurement employed by physicists to indicate the likelihood that a specific nuclear reaction will or will not take place.

The answer that emerged when that approximate figure was used, if such a volume of uranium-235 could somehow be produced, left Frisch reeling.

> To my amazement, it was very much smaller than I expected; it was not a matter of tons, it was something like a pound or two. I discussed the result with Peierls at once. I had worked out the possible efficiency of my separation system with the help of Clusius's formula, and we came to the conclusion that with something like 100,000 similar separation tubes one might produce a pound of

reasonably pure uranium-235 in a modest time, measured in weeks. At that point we stared at each other and realised that an atomic bomb might after all be possible.[17]

Not only possible. If the methodology and the mathematics employed by two 'enemy aliens' using reclaimed laboratory space at a university up to its neck in top-secret radar work had found the key to the most destructive force humankind could conjure, then surely Germany's brilliant physicists gathered at Berlin's Kaiser Wilhelm Institute were on the same path. If they hadn't already reached the destination.

Gripped by intense curiosity and surging fear, Frisch and Peierls went through more calculations to prove their astonishing theory. It was also necessary to project how much energy might be released in the fractions of a second it would take such a chain reaction to unfold. As Peierls filled in the equation by hand, the pair's astonishment was compounded.

'A rough estimate – on the back of the proverbial envelope – showed that a substantial fraction of the uranium would be split, and that therefore the energy release would be the equivalent of thousands of tons of ordinary explosive,' Peierls would explain.

'We were quite staggered by these results … As a weapon, it would be so devastating that, from a military point of view, it would be worth the effort of setting up a plant. In a classical understatement, we said to ourselves "even if this plant costs as much as a battleship, it would be worth having".'[18]

As the magnitude of their find washed over them, Frisch – a German national by dint of Austria's annexation, and already fearful that he might be bound for a British internment camp – stammered: 'shouldn't someone know about this?'

That someone was their academic superior, Mark Oliphant.

* * *

The professor listened in wide-eyed silence as the pair detailed their research, and nodded in schoolmasterly admiration as they presented their sound conclusions. He could see no initial glaring faults with the science, even if the theory tested him, and he instructed them to write it all down in a comprehensive report. Observing strictest secrecy, of course.

Such was the paranoia over possible leakage of their revelations that the men deemed the task of committing them to print too sensitive to be entrusted to the university's administrative staff. Instead, Peierls employed his basic typing skills and took on stenographic duties.

The pair worked from Peierls's small office in the single-level Nuffield Building, and with the door locked tight, the only relief from spring's first burst of warmth came by opening the window, which faced onto a section of garden between Birmingham's red-brick edifices.

Suddenly, as they discussed details that might realistically shape the ongoing war, a face appeared at the window, having arrived with neither audible warning nor logical reason. Unless – as the speechless pair hunched over the metal typewriter immediately suspected – it was an act of espionage.

'However, the "eavesdropper" was a lab technician who had planted some tomatoes along the south wall of the building, and was tending them in a spare moment,' Peierls would later recount. 'He had moved along bending over the plants and straightened up by our window. He had of course not paid any attention to what we were saying.'[19]

The five foolscap pages of single-spaced text, punctuated by hand-drawn annotations where the demands of algebraic equations exceeded Peierls's secretarial skills, were completed with a single carbon copy. They were then slipped inside a blank envelope, for hand delivery to Oliphant's office.

* * *

There, in the opening week of March 1940 and behind the polished oak door set within a floor-to-high-ceiling timber wall, Oliphant became the first (apart from its authors) to read one of the war's most influential documents. History would subsequently know it as the Frisch–Peierls Memorandum. Upon reaching the end, he set it down, gazed for a few moments through the lead-lined windows towards the physics department car park, then picked it up to begin rereading.

The report had been prepared in two parts. The first, entitled 'On the Construction of a "Super-Bomb" Based on a Nuclear Chain Reaction in Uranium', contained the technicalities, and outlined the premise of their calculations. Among the thorough explanations of uranium fission, the process for isotope separation, and the acknowledgment that some of the assumptions were based on guesswork because crucial data remained unknown, there were phrases that fairly leaped at Oliphant from the page.

'If the reaction proceeds until most of the uranium is used up, temperatures of the order of 10^{10} degrees and pressure of about 10^{13} atmospheres are produced.' In lay terms, a fireball similar in intensity to the sun's interior, and a shockwave greater than the force exerted at the earth's core, where metal is liquefied. And then: 'The energy liberated by a 5kg bomb would be equivalent to that of several thousand tons of dynamite.'[20]

The second part bore the heading 'The Properties of a Radioactive "Super-Bomb"', and was even more graphic. It was a less technical document, outlining the men's conclusions and detailing the effects such a device would likely bring. Nobody who read those two pages could claim doubt as to what the development of an atomic bomb would mean for humanity.

> The blast from such an explosion would destroy life in a wide area. The size of this area is difficult to estimate, but it will probably cover the centre of a big city. In addition, some part of the energy set free by the bomb goes to produce radioactive substances, and these will emit very powerful and dangerous radiations.
>
> The effects of these radiations is greatest immediately after the explosion, but it decays only gradually and even for days after the explosion any person entering the affected area will be killed. Some of this radioactivity will be carried along with the wind and will spread the contamination; several miles downwind this may kill people.[21]

The document went on to explain how the bomb could be constructed, with the concentrated uranium manufactured in two sub-critical parts – each of insufficient mass to enable fission – that would be kept separated 'by a few inches' during transportation to the proposed target, to avoid premature detonation. A mechanism within the bomb would then force the pieces together at the desired drop point, to set loose the chain reaction.

The report also presented five conclusions, accompanied by a caveat that, as scientists, neither Frisch nor Peierls considered himself qualified to pass comment on the strategic value of such a weapon. However, they did reiterate that while the potential of such a 'super-bomb' would make it 'practically irresistible' because no known material or structure could withstand it, it would also bring far-reaching after-effects.

The authors noted that although they held no evidence to suggest that any other scientists had reached this discovery, it was 'quite conceivable' that Germany was closing in on the same conclusions. And given that there existed no effective shelter from the destruction it would bring, the only plausible form of defence would be to threaten retaliation with a similar weapon.

'Since the separation of the necessary amount of uranium is, in the most favourable circumstances, a matter of several months, it would obviously be too late to start production when such a bomb is known to be in the hands of Germany, and the matter seems, therefore, very urgent.'[22]

If the spectre of this hideous tool under the control of Hitler's Germany did not darken the soul, then the suggestion that Britain might require specialist squads armed with measuring instruments to detect the location and toxicity of invisible radiation clouds if such a bomb were launched surely did.

The memorandum even considered, in passing, the moral debate that would rage from that day onward: whether Britain might consider such a weapon an unpalatable option, given that its impact could not be limited to purely military targets.

> Owing to the spread of radioactive substances with the wind, the bomb could probably not be used without killing large numbers of civilians, and this may make it unsuitable as a weapon for use by this country. (Use as a depth charge near a naval base suggests itself, but even there it is likely that it would cause great loss of civilian life by flooding and by the radioactive radiations.)[23]

In the decades of debate that followed the bombing of Hiroshima, Otto Frisch and Rudolf Peierls would often be asked why – upon establishing the heinous consequences of the weapon they had imagined – they did not immediately abandon their investigations.

Frisch's response would be unequivocal, and the same as Peierls's and Oliphant's when that ethical question was posed to them. 'The answer is very simple. We were at war, and the idea was reasonably obvious; very probably some German scientists had had the same idea and were working on it.'[24] As Peierls put it fifty years after the war's end: 'The thought of this weapon exclusively in Hitler's hands was a nightmare.'[25]

* * *

Oliphant stared at the pages, which were already turning soft in his clammy fingers. To the best of his knowledge, and as history would subsequently attest, he was one of three people on the planet at that moment who understood that the power of hell could at some point feasibly be delivered from the sky, to any chosen place on earth. He had quizzed Frisch and Peierls about their conclusions, and his questions were comprehensively addressed in the document he held. He also understood that if the vision of destruction outlined to him were to be realised during the conflict that continued to escalate across the Channel, then there were perhaps no better hands in which that shocking memorandum might rest.

Certainly, there was little more that 'enemy aliens' Frisch and Peierls could achieve with it. By contrast, Oliphant knew the hierarchy of Britain's war machinery, and how to get wheels moving within it. His conscience was also conflicted with the realisation that it was the partnership he had effectively forged – through his offer of safety from persecution – that had precipitated the five-page global death warrant now resting on his desk.

Oliphant had previously shown few qualms about engaging in a bit of interventionist obstruction of bureaucracy, so he might conceivably have simply returned the memorandum to its authors. Or it could have been slid quietly among other papers in a bookcase.

Instead, sitting silent and alone, at a junction where history and humanity diverged, Oliphant removed a sheet of notepaper from the top drawer of his desk and began writing a covering letter that his trusted secretary, Miss Hytch, would later type.

> I have considered these suggestions in some detail and have had considerable discussion with the authors ... with the result that I am convinced that the whole thing must be taken rather seriously, if only to make sure that the other side are not occupied in the production of such a bomb at the present time.
>
> In fact, I view the matter so seriously that I feel that immediate steps should be taken to consult with the necessary authorities concerning the possibilities of palliative measures if such a bomb should be used ... I should be very grateful if you should bring this matter to the attention of those who should be informed, though I think there are two reasons why considerable secrecy should be observed.
>
> Firstly, if we should tackle the manufacture of such a bomb ourselves a great deal of the effect it might produce would be lost if knowledge of its manufacture should be available beforehand.
>
> Secondly, if the enemy are preparing such a bomb, which is not unlikely, we should endeavour to obtain information to this effect through our Secret Service without revealing that we ourselves are aware of the possibility.
>
> I hope you will not think this a purely hare-brained scheme. It may well turn out to be impracticable, but in any case it is put forward with sincerity by Frisch and Peierls, and with considerable belief by myself.[26]

The finished letter was then affixed to the front of the five pages of typed text, and sealed inside an envelope marked 'Top Secret'. Oliphant then arranged its urgent delivery to the man he knew offered the most pragmatic hope of ensuring an atomic bomb became a reality.

It arrived on the desk of Henry Tizard on the morning of Tuesday, 19 March 1940.

18

MAUD

Birmingham and London, 1940 to 1941

The one unfamiliar face among the four that greeted Mark Oliphant when he arrived at Burlington House, mid-afternoon on 10 April 1940, confirmed the seriousness of the situation he had helped precipitate. Three weeks had elapsed since the Frisch–Peierls Memorandum was forwarded to Henry Tizard, whose response followed the best bureaucratic precedents – albeit with a rare undertone of urgency.

'What I should like,' Tizard wrote to Oliphant in response to the confronting document, 'would be to have quite a small committee to sit soon to advise what ought to be done, who should do it, and where it should be done.'[1]

The first gathering of this select group took place beneath a portrait of Isaac Newton in the main committee room of Burlington House, where the Royal Society had met for almost a century. Oliphant knew the room from previous visits, including the first, three years earlier, when he had been inducted as a fellow. He also understood that the meeting chamber was set sufficiently deep within the cobbled Annenberg Courtyard to ensure that noise from bustling Piccadilly, immediately beyond the building's grand Palladian façade, would not intrude upon deliberations.

Most of the attendees at the inaugural sitting of this hand-picked panel lent it the air of a Cavendish Laboratory reunion. There was Cockcroft,

Philip Moon and committee chair George Thomson, a Nobel Prize winning physicist and son of Rutherford's predecessor J.J. Thomson. The fourth figure, a guest dressed immaculately in a tailored dark suit who stared intently from behind heavy-rimmed glasses, was introduced as Lieutenant Jacques Allier, a Parisian banking executive recently returned from a high-stakes French intelligence mission in Norway.

Oliphant had spent much of that morning's train journey from Birmingham reading the latest news reports from Scandinavia. There were pages dedicated to the previous day's dire development, Germany's invasion of Denmark and Norway, which had shattered vain hopes that the Phoney War might yet yield a fragile peace. With that fantasy dashed, the threat to Britain was heightened. But Oliphant's more immediate concern lay with the welfare of his friend Niels Bohr – whose mother's Jewish family was prominent in Danish banking and political circles – now trapped in occupied Copenhagen.

The object of Allier's daring assignment into dangerous territory had already been explained to Oliphant. The Frenchman had been sent to secure all known stockpiles – 185 kilograms, successfully spirited to Paris via Amsterdam, Edinburgh and London – of heavy water: the source of the deuterons so coveted in Oliphant's earlier disintegration experiments, and a known moderating agent for nuclear fission.

It had been established that slowing neutrons in a fission reaction prolonged the time they spent in proximity to nuclei, thereby increasing the chances of 'capture' and the resulting atomic split. Research conducted in the United States would later establish that pure graphite served that purpose more effectively than heavy water. Yet the interest that Hitler's Germany had shown in the world's only plant producing heavy water in industrial quantity – the Norsk Hydro-Elektrisk electrochemical facility near Rjukan, in mountainous terrain west of Oslo – indicated that the Nazis were indeed on the uranium trail. This revelation was made more troubling with the overnight news that Norway was now under German control.

Allier had arrived in London a day earlier from Paris, where Frédéric Joliot-Curie had recently made significant progress in fission experiments using heavy water. Joliot-Curie had published a paper containing further details about the prospects of a self-perpetuating chain reaction in uranium. As a result, Oliphant's former Cavendish research partner Paul Harteck – who had returned to his native Germany soon after the pair's tritium discovery using heavy water in 1934 – had written urgently to the German War Office.

Harteck drew attention to 'the newest development in nuclear physics which ... will probably make it possible to produce an explosive many orders of magnitude more powerful than the conventional ones ... That country which first makes use of it has an unsurpassable advantage over the others'.[2]

Allier was only present for the first part of that initial London meeting. He informed the ashen-faced group about France's concerns regarding the measures Germany was taking to obtain information about Joliot-Curie's work, as well as their hopes to corner the world's supply of heavy water. Then, having urged that no details of his statement be shared beyond the walls of Burlington House, he was taken to see Tizard. He provided him with a list of all known German scientists likely to be involved in any nuclear research program.

The committee's next meeting would be held a fortnight hence, by which time James Chadwick would join its membership. Before long, a technical sub-committee would also be commissioned to examine practical rather than policy matters, and the main committee would grow to include Patrick Blackett.

Tizard believed it best not to pre-empt the committee's findings, yet he remained unconvinced as to the imminent prospect of a uranium bomb. After only two meetings of the 'Thomson Committee' (as the group had become unofficially known), Tizard outlined his thoughts in a letter to the War Cabinet Secretariat's assistant secretary, Wing Commander William Elliott.

'I adhere to the view that uranium disintegration is not in the least likely to be of military importance in this war. On the other hand certain physicists say that the controlled disintegration is a scientific possibility, and the French are excited about it – I think unnecessarily.'[3]

The impetus for discerning whether or not a bomb could be successfully translated from a five-page memo to a massive manufacturing facility was therefore driven largely by the men of science – of which Mark Oliphant was among the most active.

This was certainly not because they were hellbent on producing the ultimate weapon. Nevertheless, the information coming from intelligence operatives such as Allier pointed unmistakably to Germany's atomic ambitions. Given Hitler's disregard for life when pursuing his ideological obsessions, the fear that he might achieve atomic weapons capability was exacerbated by the terrifying likelihood that he would then unleash those weapons on targets that brought maximum devastation. Beating Germany to completion of the bomb seemed the only viable means of avoiding that doomsday scenario.

Naïvely, many of the scientists believed that if the theory were proven, there would be no need to deploy the bomb in combat. They imagined that a mere demonstration of its breathtaking power would end the contemporary conflict, and likely quell any sane leader's appetite to pursue war in the future.

However, the main rationale that drove Oliphant and his ilk at that time was ingrained professional curiosity. A complex question of scientific inquiry had been posed, and their trained instinct was to work tirelessly until it was answered.

* * *

So significant was the volume of research that would be needed in such urgent circumstances that the laboratory work was split among four of England's foremost universities. At Cambridge, the properties of the recently discovered element plutonium would be examined to gauge its potential as a bomb source, along with experiments on the role played by heavy water. Under Chadwick's direction at Liverpool, testing was undertaken to ascertain whether the thermal diffusion method that had been initially employed by Frisch (using the Clusius apparatus) offered the most efficient method for separating the required quantities of the uranium-235 isotope. Researchers at Oxford University were also assigned that task.

The experimental data being collected and submitted by these laboratories was sent to Birmingham, where it was studied by Peierls.

Oliphant's Birmingham facilities would also be engaged in exploring the optimum method of separating uranium-235 isotopes, but to do so he needed to identify a researcher capable of quickly coming to grips with the subject matter.

Of course, the ideal candidate already existed among his laboratory staff. However, Otto Frisch remained an enemy alien and was therefore disqualified from joining the committee, or even being briefed on its discussions. Frisch would eventually receive clearance to continue his separation experiments, but Oliphant was pointedly instructed to impress upon him the iron-clad confidentiality of his work.

To help progress the cause, Oliphant also set about converting the magnet of his cyclotron, still unfinished due to the priority placed on ongoing radar research, for use as an isotope separation system. As he had explained to colleagues at Birmingham during a lunchtime lecture, the battle against Nazi Germany loomed as 'a physicists' war'.

But sourcing further stocks of uranium so Frisch could carry out his investigations would prove a challenge, even for someone with Oliphant's talent for persuasion and, where necessary, coercion. He secured from British manufacturing giant Imperial Chemical Industries a few grams of uranium hexafluoride, a white solid that sublimes at moderate heat to form a highly toxic gas, and that was reviled in laboratories because it corroded virtually every surface it touched.

With uranium supplies so strictly controlled and the opportunities to separate meaningful volumes of the 235 isotope just as scarce, Frisch was compelled to search for alternative methods of ascertaining vital information on uranium's atomic cross-sections.

In keeping with Bohr's theory, he surmised that if natural uranium was bombarded with slow neutrons then fission would not occur in the prevalent 238 isotope. Consequently, any observable reaction must be occurring within the rare uranium 235. The difficulty with pursuing that mode of attack was that radon gas – acknowledged as the premium source of the neutrons that were then fired at the target as gamma radiation – was as difficult to procure in wartime Britain as was uranium. Unless, Frisch suspected, one set Oliphant on the trail with his gift for problem-solving.

Oliphant knew that Manchester Hospital maintained a supply of radon gas for its nuclear medicine needs and that radon, despite its half-life of just a few days, invariably replenished itself from a base supply of radium. Due to the imminent threat of air attack, however, the hospital had evacuated its radium stocks – of which there was barely 100 grams across the whole of Britain – for safekeeping to the Blue John Cavern. This was a network of subterranean shafts and passageways among the barren moors and glowering uplands of Derbyshire's windswept Peak District. The fluorescent blue stone that gave the landmark its name had been mined in the area for centuries.

Oliphant negotiated with Manchester Hospital for his Austrian researcher to gain access to the remote cave site, set more than three kilometres from the nearest village. The venture required Frisch to travel by train to Manchester, where he was met by a driver who took him the remaining forty kilometres over the gritstone plateaus and peat-rich earth of Dark Peak. His mission was to collect a 'seed' – a delicate, sealed ampule – of the known carcinogen, then transport it by the same means back to Birmingham.

Armed only with a briefcase and mild anxiety, Frisch arrived at the entrance to the Blue John's wet and gloomy subterranean chambers, which were cut into the coal-black rock and surrounded by heather-fringed grasslands.

> Down I went over slippery ladders and through narrow, muddy passages to a slightly larger cavity where, incongruously, there was a laboratory table with a lot of glassware on it, bulbs and stubs and stopcocks, rather like the equipment I had used in Hamburg. That was the plant for 'milking' the radium, for extracting the radon and compressing it into a small glass capillary, no longer than half an inch.
>
> At Oliphant's request the radium had not been milked for a whole week so that a large amount of radon had accumulated. Less than an hour later, when the local technician had done the work for me, I walked out with my little suitcase containing a heavy block of lead at the centre of which was this tiny capsule full of radon, equivalent in radiation to about three-quarters of a gram of radium.

Any safety officer would shudder at the thought that I walked out with that thing, protected by only a couple of inches of lead, and that I travelled within a few inches of that radiation source first by car and then by train. Today that would be considered an unacceptable radiation hazard both to myself and to other people in the compartment.[4]

One danger Frisch was unlikely to encounter on his day-long trip, which then led to thirty-six hours of uninterrupted experiments at the Birmingham laboratory, was having the true nature of his mission exposed. He had taken to heart Oliphant's security edict, and any information that flowed between members of the Thomson Committee and others on the periphery, including Frisch and Peierls, was deemed so sensitive that labelling such communiqués 'Top Secret' would only heighten suspicion. Instead, Peierls obtained a rubber stamp cut with a single star, which he would impress upon every confidential document ferried between the committee and the 'aliens'. Those he considered to be 'Top, Top Secret' received two stars.

* * *

It wasn't only the committee's correspondence that required rigorous security measures. The group's name also needed to be changed, given that Thomson's involvement surely identified it as pertaining to fission, since he had specialised in physics at London's Imperial College before the war.

By August 1940, it had adopted the title 'MAUD Committee', which some maintained was an acronym for Ministry of Aircraft (Production) Uranium Development Committee. Or possibly Military Application of Uranium Detonation. However, given the lengths taken to obscure the committee's purpose, these would have represented clumsily obvious codenames.

The alternative explanation is much more quixotic, and preferred by a number of the committee's members in their post-war memoirs.

It harked back to a cryptic telegram sent by Frisch's aunt, Lise Meitner, to the head of the physics department at London's King's

College during the aftermath of Germany's occupation of Denmark. This was a time when concern ran high in the global science community as to Niels Bohr's whereabouts and safety. In the hope that her update on Bohr's welfare might be disseminated to his many friends and former colleagues in Britain, Meitner cabled in economic shorthand 'Met Niels and Margarethe [Bohr's wife] recently but unhappy about recent events. Please inform Cockcroft and Maud Ray Kent. Meitner.'[5]

While confirmation that Bohr and his wife were safe – though unable to communicate directly – was gratefully received, the reference to 'Maud Ray Kent' was inexplicable. It therefore set minds racing as to the encrypted message the Dane was seemingly trying to send.

Cockcroft felt it referred to the unpalatable reality that the Nazis had successfully developed some sort of death ray. Intelligence officers called in to try to decipher the oblique meaning thought it was more likely a warning about Germany's progress in the race for an atomic bomb. By replacing the 'y' in 'Maud Ray Kent' with 'i', it became an anagram of 'radium taken'.

Concern grew to the point where Ralph Fowler, Rutherford's son-in-law, wrote to Tizard pointing out that while some of the theories floated were fanciful, the premise underpinning them was 'sufficiently reasonable to make one worry'.[6] Ever the pragmatist, Oliphant ventured that the whole matter could be readily resolved by having the British Embassy in Stockholm, the city where Meitner continued to live in exile, contact her directly for clarification.

That route was not pursued, however, and it was not until three years after the MAUD Committee earned its title that Bohr was finally smuggled out of Denmark and the truth was revealed. This happened when Bohr inquired as to whether his original message had been passed on, as requested, to his family's much-loved governess during their earlier residence in Britain – Miss Maud Ray, who had since relocated to the south-east county of Kent. Her exact address, rather like the meaning that gave rise to myriad conspiracy theories, had been lost in transmission.

* * *

The MAUD Committee's heaviest workload coincided with the Germans' most sustained bombing attacks. From the late summer of 1940 through to the final weeks of spring the following year, around 30,000 tons of explosives were dropped on British targets stretching from Cardiff in South Wales to Scotland's western port of Glasgow. The number of dead topped 40,000, many of them civilians. Hitler's plan to pummel Britain into surrender meant residential tenements became as vulnerable as munitions factories.

The village of Barnt Green – where Peto's red-brick walls now echoed with emptiness and remained bitterly cold due to the shortage of available coal – might have been spared the nightly threat of Luftwaffe bombers. Birmingham, however, remained a round-the-clock war zone. The network of industrial-era transportation canals radiating from the city's centre provided a handy navigational aid for Luftwaffe pilots, and fires blazed across the city on a nightly basis.

Amid the howl of incoming explosives and bursts of shrapnel from anti-aircraft shells, the distinctive purple livery of vans belonging to the Cadbury family's eponymous chocolate company, founded in Birmingham, became a regular sight. They would arrive at buildings where firefighters battled flames, and young women wearing steel helmets would alight to administer fortifying mugs of steaming cocoa.

As the Nazi air offensive worsened, the university came under direct threat. In mid-1940, in the first weeks of the Battle of Britain, an incendiary bomb exploded within a metre of the front gates. When another penetrated the supposedly fireproof roof of Oliphant's prized Nuffield Laboratory and set the rafters ablaze, the campus fire brigade tore out the ceiling and extinguished the flames while ensuring that valuable cyclotron equipment within was spared from water damage.

Staff were assigned to teams that took turns in maintaining night-time vigils. Between bouts of fitful rest on a makeshift bed in the laboratory, Oliphant would keep watch through the porthole windows of the Poynting Building's bulwark façade. From there, he would observe the orange light through the billowing smoke that cast an apocalyptic glow upon the horizon.

> One night, the air-raid sirens alerted me. Half asleep I pulled on a dressing gown and made for the door to see what was happening. I walked straight into the glass door of the library. This shattered with a tremendous noise, cutting my eyebrow so that blood poured down my face. Putting my hands up, I thought 'that bomb got me! I'm finished!' Then I was somewhat disappointed to find that I had survived.[7]

Later in the war, Oliphant arrived at work one morning to find that a 500-pound (225-kilogram) bomb had slammed through the Poynting Building's roof and into his first-floor office, where it lay unexploded behind a bookcase. The delicate defusing required army experts, who forbade the professor from entering the room, either to watch them operate or to access his files. 'I was annoyed, for I wanted urgently some correspondence there about a visit to America, but they were adamant.'[8]

At the height of the Blitz, Oliphant travelled to London at least once a month when the MAUD men convened at Burlington House. The timing of the committee's sixth meeting in mid-September 1940 was especially traumatic, as London had sustained relentless bombing raids every night for two weeks. After months of targeting RAF airfields and radar installations, Hitler had decided that London would wear the assault designed to bring Britain to its knees.

Pulitzer Prize winning war correspondent Ernie Pyle later wrote movingly of that time:

> You have all seen big fires, but I doubt if you have ever seen the whole horizon of a city lined with great fires – scores of them, perhaps hundreds. There was something inspiring just in the awful savagery of it.
>
> Little fires grew into big ones even as we watched. Big ones died down under the firemen's valour, only to break out again later. About every two minutes a new wave of planes would be over. The motors seemed to grind rather than roar, and to have an angry pulsation, like a bee buzzing in blind fury.[9]

So Oliphant was understandably edgy in the wake of that September meeting, as his return train to Birmingham edged out of London's Euston Station after nightfall.

Minutes into the journey, the train came to a dead halt. The scream of air-raid sirens piercing the constant low rumble of approaching bombers that played to a background chorus of exploding shells drew Oliphant to the blacked-out window of his carriage. He lowered it sufficiently to crane his head towards the pyrotechnics that were punching holes in the night sky. His curiosity was driven partly by the chance to watch how effectively London's anti-aircraft guns operated, and to assess how they might benefit from some form of radar guidance.

As horns wailed around him, the view from the railway bridge on which his train had come to rest was both daunting and dazzling. Rounds of heavy artillery blazed against the blackness, and the glow cast everything in an amber pall.

Suddenly a huge blast from beneath the train rocked his carriage and sent him crashing to the dusty floor. He was convinced, once more, that he had fallen victim to an enemy bomb. Again, however, after a few moments of stunned disorientation, he was able to clamber to his feet, unhurt. The source of the explosion was then revealed to be a battery of guns installed not 100 metres away, beneath the very same bridge.

It was during his regular journeys to the capital, and the hours he spent caged within rail carriages or walking to and from his accommodation at the Athenaeum Club on Pall Mall, that Oliphant's resolve to employ any means available to end the war gained steel. 'It was the … hatred of the Hitler regime that really drove me,' Oliphant later recalled of his wartime work.[10]

> There were times during the Battle of Britain when, following massive air raids in London, seeing the havoc among people and buildings as I walked the streets in the grey light of early morning, I felt hatred of the enemy welling up within me, so that I was nauseated and longed for retribution. Perhaps it was because, as a student, I had so liked and admired the Germans, that my revulsion

was so powerful. Now … I realise that we know very little about such deep human emotions, from which flows the desire to kill.[11]

* * *

At the heart of those MAUD Committee meetings held during the relentless Blitz of September 1940 and beyond were two questions.

One was how to determine the most effective means of separating a critical volume of the uranium-235 isotope. The thermal diffusion technique employing Clusius tubes initially favoured by Frisch had been deemed too time consuming. Thus the favoured method became 'ordinary' diffusion, which required the toxic gas uranium hexafluoride to be forced through tiny holes or slits punched in a metal plate, so it could separate the lighter isotopes. However, the uncertainties attached to this method meant much further experimentation and detailed calculations were required. Oliphant took it upon himself to assign Peierls to the case.

The other significant query was the precise size of the critical mass required to sustain a chain reaction in a bomb. It was this need to understand more about the mysterious properties of uranium isotopes in gaseous forms that effectively sidelined Oliphant from the initial MAUD Committee – now officially known as the MAUD Policy Committee – in early 1941. In order to keep that panel to a manageable number, membership was limited to one representative per institution or department. Birmingham's man would be chemistry professor Norman Haworth, who could lead the examination of uranium's chemical mysteries.

Further calculations had already shown that the likely quantities advanced by Frisch and Peierls in their famous memorandum were significantly short of what would be needed. Instead of 'a pound or two' of uranium-235 that had been separated and therefore enriched, the core of a bomb was likely to require around twenty-five times that amount. This, in turn, meant that the time necessary to refine it, and the cost of doing so, also increased exponentially.

Had the pair's original 'back of the envelope' sums been informed by knowledge subsequently available to the MAUD Committee, it is quite likely that Tizard's early doubts would have seen the bomb project shelved – at least until after Britain had dealt with the threat of German invasion.

* * *

As a result of the committee reshuffle, Oliphant's direct input was restricted to matters of logistics and process. However, he was a potent presence when the subsidiary MAUD Technical Committee convened in Burlington House at 11am on 2 July 1941, to finalise the report they would submit via Tizard to the Scientific Advisory Committee of the War Cabinet, chaired by Lord Hankey.

Along with other MAUD members, Oliphant had received a draft version of Thomson's final report a week earlier, and asked if he saw the need for any major amendments. When the group gathered that morning, Oliphant pointed out that the intricate detail contained in the lengthy document was unlikely to be understood by anyone beyond those walls. If production of a bomb were to be pursued by military and political leaders, they would not be persuaded by pages of scientific jargon. They would need a simple over-arching summary of whether it would work, and how that might be achieved.

The first comprehensive road map to the development of an atomic bomb was submitted to Tizard on 15 July 1941. In its final iteration, the MAUD Report ran to more than thirty pages of dense text, a scattering of mathematical formulae and tables, and a diagram of the proposed isotope separation plant. It also listed technical details on the properties of uranium, the critical mass, projected size and fusing options for the weapon, estimates of the damage it could inflict and a list of production problems the process faced. Finally, the document featured multiple appendices addressing the engineering issues that such a vast undertaking would encounter.

Its explicit premise, however, was spelled out in the three-page summary for which Oliphant had successfully argued. The main report began by acknowledging that, at the start of the process of inquiry, the committee members had been more inclined to scepticism than belief.

Over the course of fifteen months, however, that view had evolved to the point where they were convinced that a bomb was indeed feasible.

As the report's opening summary noted: 'It will be possible to make an effective uranium bomb which, containing some 25lbs [eleven kilograms] of active material, would be equivalent as regards destructive effect to 1800 tons of TNT and would also release large quantities of radioactive substances, which would make places near to where the bomb exploded dangerous to human life for a long period.'[12]

As it happened, 1800 tons of TNT was equivalent to the total explosives payload that Germany had dropped on London during the nine-month Blitz. It had ended in May 1941, just two months before the MAUD Report was finalised, at a cost of more than 20,000 civilian lives.[13]

The specifications outlined for the bomb were almost as daunting as the projections of its power. The estimated cost of a separation plant able to produce one kilogram of uranium-235 per day – the yield needed to fuel three bombs per month – would be approximately £5 million (almost £250 million today). The time needed to build a manufacturing plant for uranium separation and bomb fusing would be just over two years, with an estimated completion date of late 1943. The plant would comprise around 1900 individual isotope separation units, and occupy an industrial footprint of almost 700 square metres.

The raw expense appeared prohibitive for a nation still engaged in a fight for its very survival, even if Hitler's attention had recently turned to Russia. Yet the report calculated that a fully operational production facility should be capable of turning out sufficient enriched uranium fuel for thirty-six bombs per year, at a cost per weapon of around £236,000. By comparison, 1800 tons of TNT in regular bomb form represented an outlay of £392,000. Putting aside the huge establishment cost, production of a British atomic bomb in per-weapon terms made economic sense.

It would be the fusing mechanism, needed to slam together the two sub-critical sections of enriched uranium at a speed of around 6000 feet per second, that would account for most of the bomb's mass. However, the finished device was not expected to weigh more than a ton, and would therefore be within the carrying capacity of a modern bomber aircraft.

Items of concern included the likely need for the bomb to be deployed using a parachute, to allow the transporting plane time to clear the huge blast zone before detonation. Another issue was that, because a fission chain reaction could not occur without the calculated critical amount of uranium-235, 'the main principle cannot be tested on a small scale'.[14] There was no scope for a concept design to be proven via an atomic hand grenade.

Not that those mitigating factors were canvassed in the final summation, which pronounced that the MAUD Policy Committee 'considers that the scheme for a uranium bomb is practicable and likely to lead to decisive results in the war. It recommends that this work be continued on the highest priority and on the increasing scale necessary to obtain the weapon in the shortest possible time.'[15]

In addition to the bullish forecasts about the infrastructure and the practicalities of building a bomb, Tizard's attention was drawn to an item addressed in the summary and detailed further in the body of the document.

The report mentioned that British intelligence had already learned that, since the Germans had occupied Norway, they had actively increased production levels of heavy water at the Norsk Hydro plant.

A daring sabotage, planned in Britain and involving Norwegian commandoes, would be launched on the fortress-like site in late 1942 and, when that proved a disaster, reprised with greater success the following February. The destruction of this plant, combined with the stockpile that Lieutenant Allier had helped remove from Norway under the Germans' noses — and which ultimately found its way to back to Britain following the fall of France — meant that the Nazis' access to the rare commodity had been stopped, for the time being. The plant would be repaired, however, and heavy water production resumed. This would require a further, final mission on Norway's Lake Tinnsjø two years later.

Upon receiving the report in mid-1941, Tizard understood that if German scientists were trying to get hold of the moderating agent, it could only be for use in nuclear fission work. The MAUD

Report had explained that while heavy water was valuable for use in fission experiments designed to produce nuclear energy, its value in the production of a bomb was limited. Once the Germans realised this, and accepted that pure graphite worked far more efficiently as a moderator, Tizard believed Britain's competitive edge in the bomb race would be lost.

* * *

From Lord Hankey of the Scientific Advisory Committee, the MAUD Report progressed to Lord Cherwell, Winston Churchill's hugely influential scientific advisor. Cherwell then drafted a note to the Prime Minister on 27 August 1941, and his advice was as resolute as Tizard's initial view had been equivocal.

'I am quite clear that we must go forward,' he counselled Churchill. 'It would be unfortunate if we let the Germans develop a process ahead of us by means of which they could defeat us in war or reverse the verdict after they had been defeated.'[16]

Three days after receiving the communiqué – and less than eighteen months from the time when Oliphant had sat alone in his office in Birmingham, absorbing the news that a uranium fission bomb might be possible – Churchill drafted a historic memo to his military chiefs of staff.

'Although personally I am quite content with the existing explosives,' he wrote, 'I feel that we must not stand in the path of improvement, and I therefore think that action should be taken in the sense proposed by Lord Cherwell.'[17]

There was no higher British administrative authority than Churchill's. The long march to the development of an atomic bomb would proceed.

But having effectively set in motion that chain of occurrences, Oliphant was taken aback by the next development. The escalation of the project to government's top level had meant the formation of another committee – this time a consultative council, which decided that the bomb development work should be entrusted, in turn, to a new division within the Department of Scientific and Industrial Research.

For reasons of opacity, that body would be known as the Directorate of Tube Alloys. The man chosen to oversee the new entity, which would replace the MAUD committees, was Wallace Akers, Director of Research at Imperial Chemical Industries (ICI). Akers had initially proposed the body be called 'Tank Alloys', to signify that its importance to the military effort corresponded with that of the heavy artillery vehicle. However, the minister responsible for the project, Sir John Anderson, pointed out that the bomb's success might well signal the end of the tank's pre-eminence. Therefore, a more enduring symbol of armaments manufacture – the simple tube, with its many applications – was more apt.

Whatever the reasoning behind the name, the decision to disband MAUD and place development of the bomb under the jurisdiction of an individual with clear commercial loyalties – without a hint of consultation or forewarning – annoyed many committee members.

It infuriated the more combustible Oliphant. He penned a scorching letter to the Secretary of the Department of Scientific and Industrial Research, Sir Edward Appleton, complaining bitterly that this most complicated assignment was now being managed by people utterly lacking the requisite scientific rigour, and threatening to resign his post immediately.

'I, personally, was responsible for the whole of the recrudescence of this subject and had to fight hard to get things going and to get Frisch and Peierls accepted,' Oliphant fumed. 'In the reorganisation I was left off the policy committee and I feel that the time has now come for me to sever my connection altogether. This problem is too important to be trifled with.'[18]

Writing to Chadwick, Oliphant described the appointment of Akers as 'disgraceful', because he would 'obviously ... look after the commercial interests of ICI'. In the same letter, Oliphant even went so far as to threaten to lead a mutiny of other disgruntled MAUD members and enlist them to set up a 'rival show', which would allow them to continue their research without 'a lot of interfering busybodies who know nothing whatever about the problems involved'.[19]

Despite his headstrong belief that science should proceed unhindered for its own noble sake, not bend to the whims of political or private interests, Oliphant was eventually placated by the more measured Chadwick. As his younger brothers, his university contemporaries, and even a few of his Cavendish colleagues well knew, Oliphant's flashes of indignant temper usually passed like an Adelaide summer storm: much noise and spectacular fireworks before the sun surely reappeared.

In this instance, Oliphant came to appreciate that the magnitude of the project, as detailed in the MAUD Report, required manufacturing and management expertise on a scale that lay beyond the capabilities of the scientists' collective. And he would even come to develop admiration for the skills and sensibilities of Akers.

* * *

One recommendation by the MAUD Committee that had addressed Tizard's nagging doubts about the bomb's likely success was the reference to ongoing co-operation between British scientists and their American counterparts. The report's suggestion that developmental work on isotope separation should continue on both sides of the Atlantic resonated with Tizard, who believed the entire project would be better pursued in the United States. He had admitted as much in a letter to Lord Hankey:

> I ... think it is absurd to embark on this very big and highly speculative industrial undertaking in this country with all we have to do in other ways, and I think the only sensible thing to do is to send Chadwick and Thomson to America to discuss all the results with the Americans who have been doing similar work, and on the basis of that decide whether a plant should be put up somewhere in North America.[20]

Indeed, the United States had established a uranium committee a year before MAUD first met in 1940. Across the Atlantic, the force driving investigations into uranium fission, and the possibility it might yield a 'super-bomb', was the man considered the world's foremost physicist after Rutherford's passing: Albert Einstein.

Shortly before Poland fell in 1939, Einstein had been contacted by Leo Szilard, the Hungarian-born physicist who first foresaw the possibility of a nuclear chain reaction. Szilard's spiralling concern led him to approach his friend about lending his revered name to a letter to the Queen of Belgium, formally requesting that she intervene to prevent the sale of Belgian Congo uranium to Nazi Germany. However, upon realising that Einstein's repute meant he stood a chance of getting a message directly to United States President Franklin D. Roosevelt, Szilard saw the prospect of an even bigger win. When he explained the theory behind uranium fission to Einstein, months before Frisch and Peierls began their calculations, the greatest thinker of the twentieth century exclaimed simply: 'that never occurred to me'.[21]

By the time Einstein's letter reached the White House, Poland was under German occupation. The note's text subtly urged Roosevelt to commission a group of physicists in America to explore chain reactions, but it also reflected scientific opinion as it existed in late 1939 in pointing out: 'A single bomb of this type, carried by boat or exploded in a port, might well destroy the whole port together with some of the surrounding territory. However, such bombs might very well prove to be too heavy for transportation by air.'[22]

Come the war's end, Einstein would declare this letter to be his sole contribution to the development of the atomic bomb, a project with which he is often spuriously linked. But even though it bore his weighty signature, the letter's impact was muted.

Roosevelt instructed that a three-man advisory committee on uranium be established, reporting to Lyman J. Briggs, director of the National Bureau of Standards. He proved an uninspired choice. With a four-decade background in agricultural science, Briggs was daunted by the assignment handed to the Advisory Committee on Uranium, and consequently it had made little progress.

Soon after the MAUD Committee had been set up, its members had contacted Archibald Hill, a Cambridge old boy and Nobel laureate who had recently been posted to the British Embassy in Washington. Hill was asked to establish the level of interest in nuclear fission in the United States, in order to ascertain whether a joint venture might be worth pursuing.

Hill's response in May 1940 had been as unambiguous as it was deflating.

> There is, I am informed here, no possibility within practicable range of using uranium either as a power source or as an explosive. It is not inconceivable that practical engineering applications and war uses may emerge in the end. But I am assured by American colleagues that there is no sign of them at present and that it would be a sheer waste of time for people busy with urgent matters in England to turn to uranium as a war investigation.
>
> If anything likely to be of war value emerges they will certainly give us a hint of it in good time. A large number of American physicists are working on or interested in the subject; they have excellent facilities and equipment: they are extremely well disposed towards us: and they feel that it is much better that they should be pressing on with this than that our people should be wasting their time on what is scientifically very interesting, but for present practical needs probably a wild goose chase.[23]

However, more than a year later, with Britain having been given prime ministerial approval to pursue the development of an atomic bomb, there was hope that American attitudes might have changed.

In keeping with the spirit of co-operation that had brought such success with the cavity magnetron, and had led more than 3000 British military officers and businessmen to visit the United States by mid-1941 to promote their embattled nation's cause, regular updates on the MAUD Committee's uranium deliberations had been fed to its equivalent body in America.

A copy of the MAUD Report and all its accompanying technical appendices had been provided to Dr Briggs at the uranium committee upon the report's completion in July 1941. It had been accompanied by a note requesting that it be forwarded to Roosevelt's powerful science tsar, Vannevar Bush, and to James B. Conant, head of America's peak science body, the National Defense Research Committee.

Weeks later, George Thomson – chair of the two soon to be disbanded MAUD committees – expressed surprise that no response had been received from any of the American recipients in relation to the report's seismic findings.

Around the same time, in August 1941, Oliphant was advised that he would be travelling to the United States to engage in further information-swapping on radar issues stemming from the Tizard Mission. Shortly before his departure, however, Oliphant received a letter from Thomson. It stressed that although Oliphant's principal mission was furthering the crucial trans-Atlantic radar partnership forged a year earlier, he should also make a few 'guarded inquiries' as to how the MAUD Report had been received in the United States.

It was a directive that Oliphant chose to interpret as a mandate to relentlessly push the imperative of America's involvement in the bomb project.

This campaign would prove so fervent, and ultimately persuasive, that he could later lay claim to being a founding father of the Manhattan Project that would follow.

19

'MEDDLING FOREIGNER'

United States and Birmingham, 1941

Although the accepted means of safe passage from Britain to the United States in 1941 was the Pan American clipper service through Lisbon, the Portuguese capital seethed with spies, who would surely identify such a recognisable, eminent physicist as Oliphant and thus compromise his journey's confidentiality. It was also a comparatively slow route, and time was ticking.

Instead, Oliphant took the option of an unheated, heavyweight B-24 Liberator bomber run from the RAF-operated Prestwick air base in Scotland. He was seemingly unfazed by whispers that a previous passenger had lost digits to frostbite during a similar sixteen-hour, high-altitude haul to Newfoundland on Canada's east coast. Investigations into that persistent rumour had revealed that the unfortunate victim had been an airman whose flying boots were too tight, resulting in interrupted blood flow that had cost him a couple of toes. Everything else about the enterprise, though, seemed steeped in mystery.

Civilians berthed on these clandestine flights would receive cryptic confirmation shortly before departure – simple advice that a sleeper berth had been reserved aboard the night train to Glasgow, and to alight at Kilmarnock where further instructions would be forthcoming. Passengers would then rendezvous with an unmarked vehicle that appeared at Kilmarnock railway station, and were ferried to a makeshift air terminal nearby. There they were weighed (along with their luggage), and flying suits were issued to stave off the piercing cold of their imminent journey. Oxygen masks were mandatory in the non-pressurised military plane, and Benzedrine (amphetamine) tablets were provided just in case anyone found themselves drifting off to sleep despite the freezing, unfurnished interior.

Oliphant's gravitas spared him that cattle-class indignity, however, and he was allocated a cockpit seat, which meant he was privy to the final dramatic minutes of his first trans-Atlantic flight. It was scheduled to terminate at Gander airbase on Newfoundland Island, but with fuel critically low and the flight deck's navigator afforded minimal visibility, the lumbering aircraft dropped out of low cloud to find an expanse of water immediately below, where land was meant to be. The frantic pilot found a freshly laid earthen landing strip among the marsh of waterways, where the craft put down safely.

Undeterred, Oliphant immediately commenced his onward journey, which took him to the land of plenty. After two years on the besieged island of Britain, he quickly became transfixed by the sight of fresh fruit stalls on American city streets, and public lighting that blazed after dark.

* * *

The radar business he was to pursue with the United States National Defense Research Committee took him to the Radiation Lab at MIT in Boston, to Bell Telephones in New York, and finally to Smithsonian Institution in Washington, DC. All of these bodies received Oliphant with great warmth, as befitting his standing as overlord of the department that had devised the crucial microwave apparatus.

Accompanied by Cavendish alumnus Charles Galton Darwin – now Director of the British Central Scientific Office in the United States capital – Oliphant then embarked on his unofficial mission, which would ultimately supersede his radar commitments.

The first meeting the two men would conduct on the sensitive matter of the atomic bomb was with Lyman Briggs, the man entrusted a month earlier with a copy of the final MAUD Report.

Seated in Briggs's stifling office, Oliphant quickly comprehended the sticking point. Despite his self-effacing nature and permanent half-smile, the white-haired Briggs struck his visitor as an insipid apparatchik whose lack of initiative and limited comprehension of the science were obvious. To compensate, Oliphant surmised, Briggs had developed 'a mild mania about the secrecy business',[1] which extended to hiding details from some of those directly providing the uranium committee's experimental expertise.

Oliphant's temperature was rising, and Washington's early August heat was only partially to blame for the steam that began to build beneath his sharply starched collar. For if there was a trait that riled him above most others, it was inertia. He watched Briggs roll the bowl of his empty pipe between thumb and forefinger, staring at Oliphant with heavy lidded eyes from across an expansive desk. In pride of place upon it was a tiny cube of dull metal that his agitated guest had immediately recognised as uranium.

While Briggs's uranium committee had been holding regular meetings over the past year, it became apparent that little of that time had been spent even contemplating the prospect of an atomic weapon. Oliphant's incredulity boiled over upon learning that the MAUD Report had not made it out of Briggs's keeping – or even onto his desk, for the most cursory perusal.

Oliphant would recall that 'this inarticulate and unimpressive man had put the reports in his safe and had not shown them to members of his Committee. Amazed and distressed, I reported the situation to [the report's intended recipients] Vannevar Bush and James B. Conant …'[2]

Bush and Conant had been privy to earlier discussions of the MAUD Committee, having been provided with meeting minutes since early 1940. They had even seen a draft of its report before the final iteration was submitted to Briggs. But such was the agreed need for concealment in the United States that nobody seemed prepared to discuss its findings.

They had not, however, reckoned on Oliphant's obduracy – nor his belief that confidentiality should always take a back seat to the greater good.

Thomson might have only suggested that Oliphant make a few 'guarded inquiries' upon landing in America, but Oliphant saw that engendering support in America for the MAUD Committee's findings would require the same sort of shameless hectoring he had employed to fund his Birmingham cyclotron. He would be described by physicist and science writer Jeremy Bernstein as 'like a man possessed … he simply would not be contained when it came to discussing prospects of a bomb'.[3]

He took it upon himself to dedicate the remainder of his United States visit to a relentless, one-man campaign devised to deliver a clear, unauthorised outcome: to convince all scientific peers with whom he came into contact of the dire need for America to get serious about an atomic bomb. Before it became Germany's decisive weapon.

* * *

After the unfortunate meeting with Briggs, the campaign began in earnest with Bush and Conant.

Oliphant's dinner with Conant in Washington prompted much interested listening but little enthusiasm. His subsequent twenty-minute meeting with Bush in New York was less encouraging still. 'They were not interested,' Oliphant would recall of his interactions with America's pre-eminent science administrators. 'They said "oh that's for the next war, not for this one".[4]

When these discussions met clear resistance, Oliphant took his message directly to a meeting of the uranium committee at which Conant was present, bolstered by his belief that he held immunity to discuss the potential of a bomb with any colleague he thought suitably influential.

As a newly installed member of Briggs's panel, Samuel Allison from the University of Chicago was taken aback by the confronting message laid out by the avuncular, bespectacled guest from England. Like most of his colleagues, Allison had been utterly unaware of the MAUD Committee's findings, until Oliphant's blunt presentation set him straight.

'He came to a meeting of the Uranium Committee and said "bomb" in no uncertain terms,' Allison later recalled of that autumn day. 'He told us we must concentrate every effort on the bomb and said we had no right to work on power plants or anything but the bomb. The bomb would cost twenty five million dollars, he said, and Britain didn't have the money or the manpower, so it was up to us.'[5]

Oliphant also sought out influential figures at General Electric's headquarters at Schenectady in New York State. There he met with William Coolidge, who had been commissioned by Bush two months earlier to produce a review of the uranium committee's operations. It had reiterated the findings of a previous report Bush had commissioned earlier in 1941: that a uranium bomb was a distant proposition.

Coolidge was at first delighted to welcome a visitor of Oliphant's eminence. This was, after all, a physicist bathed in the Cavendish Laboratory's glorious achievements, and whose name appeared on published papers alongside that of the late Lord Rutherford. Then, after Coolidge was regaled with information unavailable to him during the preparation of his review because it had been sitting, unopened, in Briggs's safe, the credulity of GE's acting chairman was stretched.

He wrote immediately to Frank Jewett, President of the National Academy of Sciences, the body to which both review reports had been sent.

> I was greatly interested in what Dr Oliphant told me on his recent visit to Schenectady. You may well have heard his story on the diffusion method of separating the isotopes and his prediction that only 10 kilograms of the pure 235 would be needed for the chain reaction ... He said that it had been estimated that the 10kg bomb of pure 235 would be equivalent in explosive effect to 1,000 tons of ordinary high-power explosive and could be used to destroy everything in an area over a mile in diameter ...
>
> This information, so far as I know, was not available in this country ... I think that Oliphant's story should be given serious consideration as it may make a further study of separation by diffusion look far more important in connection with national defense than the further work that we recommended ...[6]

Oliphant's relentless prosecution of the case for the atomic weapon had finally stirred some influential support. He was, however, yet to play his trump card.

* * *

By now Oliphant's kinship with Ernest Lawrence had grown beyond their shared devotion to science. A colleague who worked closely with both men observed of the pair: 'They were as alike as two peas in a pod. Oliphant knew a little more physics than Lawrence, but both were energetic developers, essentially promoters'.[7] They also felt mutual frustration at the inaction of Briggs's uranium committee, with Lawrence deriding the ponderous chairman as 'slow, conservative, methodical and accustomed to operate at peacetime government bureau tempo'.[8]

The bond between the two physicists had been galvanised at the outbreak of the war, after Lawrence's younger brother, John, sailed to England to attend a conference staged by Oliphant's physics department at Birmingham University. On the day after Germany invaded Poland, John Lawrence boarded the unarmed passenger steamer *Athenia* at Liverpool for his return voyage to the United States. When the ship was struck by a pair of German U-boat torpedoes the next evening, Oliphant became the trusted source to whom Ernest turned in an effort to establish his brother's fate, which was still unknown after days of monitoring every radio news bulletin.

Knowing well the helpless anxiety engendered by being distant from loved ones in times of trauma, Oliphant enlisted his friends in high places to establish that John Lawrence had bravely remained on board the *Athenia*, in the inky vastness of the North Atlantic, to help injured passengers and crew into rescue vessels. He had then found safe refuge in the last of the lifeboats, which made for Ireland as the stricken ship slipped forlornly to the ocean floor. On his return to the United States, John Lawrence pursued a highly successful and decorated career exploring the use of radioactivity in the treatment of cancer.

The *Athenia* incident had stirred Ernest Lawrence's will to fight the war using the full artillery of his nuclear research, in whatever capacity might be required, in the same way Oliphant's London Blitz experiences had hardened his heart against Germany.

With the war confined to Europe over the intervening years, such sentiment in the American science community had been largely confined to political and Jewish refugees such as Leo Szilard. But Oliphant well knew that Lawrence shared his own repugnance for Germany and its atomic aspirations and, given Lawrence's scientific clout on both sides of the American continent, he would be the final rallying point before Oliphant's return to Britain.

'I'll even fly from Washington to meet at a convenient time in Berkeley,'[9] Oliphant cabled Lawrence.

Despite having been in the United States for more than six weeks, on a mission that had long since met its radar objectives, Oliphant boarded another uncomfortable military flight – this time eighteen hours in a frigid DC-2, from New York to San Francisco – when he might just have happily spent that period of discomfort making his way in the opposite direction, towards home.

He was met at the airport by Lawrence's laboratory deputy, Donald Cookesy, who taxied the weary visitor to Berkeley's bayside campus. There, on the afternoon of Tuesday, 23 September 1941, one of the more critical meetings in the history of atomic weaponry took place.

Lawrence was anxious to show off his latest project, the monstrous 184-inch (4.6-metre) cyclotron taking shape at the crest of the city's Charter Hill, and after a short drive up the snaking dirt track dubbed 'Cyclotron Road', he and Oliphant alighted beneath the world's largest magnet.

With the sight and scent of eucalyptus trees – so redolent of his boyhood in the Adelaide Hills – behind him to the east, and a westward view in which the recently completed Golden Gate Bridge spanned the sparkling expanse of San Francisco Bay, Oliphant was struck by the site's aesthetic beauty as much as the proposed cyclotron's latent power. Then he snapped back to business, and the matter that had brought him there: the need for the United States to seriously embrace development of the bomb.

'We discussed the general problem, and in particular the methods which we had been considering in Britain for the separation of the isotopes of uranium,' Oliphant later wrote of the conversation with Lawrence. 'He was deeply impressed by the serious view of scientists in England that nuclear weapons were not only almost certainly possible, but that Germany might be working on the problem.'[10]

The two friends then made their way back down the hill to Lawrence's office, where Oliphant unhesitatingly reeled off a detailed description of material contained in the confidential MAUD Report.

He barely paused for breath even after they were joined by Lawrence's physics department colleague Robert Oppenheimer, whose eyes bulged wide at the news he was hearing for the first time. If Oliphant was suddenly concerned at having let slip so fundamental a secret to someone on the periphery of nuclear research, with an unknown security profile, he gave no indication.

In fact, it was Oppenheimer who politely queried the wisdom of the visitor's intemperance, and suggested Oliphant might refrain from openly canvassing a topic of such obvious military importance with which Lawrence's colleague was clearly unfamiliar. 'But that's terrible,' Oliphant forged on, unapologetic and unperturbed. 'We need you.'[11] All the while, Lawrence sat behind his desk, nodding sagely in agreement.

As Oliphant prepared to fly back to the east coast, then on to Britain, Lawrence asked him to prepare a statement summarising the MAUD Report's conclusions, for the benefit of the majority in the American science community who had been deliberately kept in the dark as to its contents.

Oliphant, who in his own official report of discussions held with Lawrence during that visit claimed his colleague had 'only come to know of its [the MAUD Committee's] existence by accident',[12] duly completed a two-page summation. It noted the MAUD Committee had considered the respective merits of pursuing a fast-neutron fission bomb with either uranium-235 or plutonium at its core. In acknowledging the case for plutonium (which was expected to undergo fission more readily than the uranium isotope), Oliphant restated the MAUD Committee's belief that 'from the point of view of the war the separation of the U-235 isotope is alone justified', even though that separation process, compared to the less cumbersome means by which plutonium could ultimately be manufactured in nuclear reactors, would be time consuming and expensive.

His precis also recorded the committee's majority view that such a project was beyond Britain's current means, and would therefore need to take place in either the United States or Canada, as a joint enterprise. Oliphant's emphatic conclusion included an admission that he carried no formal authority to propose such a view, but that, in itself, was scarcely grounds for him to stay silent.

> Finally, I would like to say that the preparation of a nuclear bomb, if this be possible, should be undertaken at once and on the very highest priority. We cannot afford to neglect even a probability, that the scheme will work successfully. Whichever nation is first to succeed in this quest will undoubtedly be master of the world. If peace were to come tomorrow it would still be necessary to obtain the answer first at all costs, for in the hands of a resentful or unscrupulous nation such power would be dangerous.
>
> While I have discussed the MAUD Committee and made very definite statements in this memorandum, it must be made perfectly clear that I speak as an individual without any official status or authority in this matter.[13]

* * *

The impact of Oliphant's unofficial United States mission was both profound and immediate. No sooner had he departed San Francisco than Lawrence telephoned Arthur Compton, a Nobel laureate and the hugely influential professor of physics at the University of Chicago. In that call, Lawrence volubly echoed Oliphant's view that the outcome of the current conflict likely rested on the race to the atomic bomb.

It would be the beginning of a confluence of events over coming weeks that saw history gain pace, hastened by a free fall of coincidence.

In the wake of Oliphant's barnstorming tour, the once tightly held findings of the MAUD Committee became an open secret in America's scientific circles. As history professor and author Ferenc Szasz would write:

> J. Robert Oppenheimer later admitted that the MAUD Report transformed the American [nuclear] program from a series of

desultory committees to a focused, concentrated effort. Historian A.J.R. Groom has estimated that the MAUD Report accelerated the American weapons program by a minimum of six months. [Nuclear historian] Margaret Gowing stated it even more forcefully: 'Without it [the MAUD Report],' she wrote, 'World War Two would almost certainly have ended before an atomic bomb was dropped.'[14]

While Mark Oliphant was central in the preparation and dissemination of both the Frisch–Peierls Memorandum and the MAUD Report, it was his tireless lobbying and impassioned badgering of his scientific brethren in the United States that would prove his most tangible contribution. It would lead Szilard to reflect on the cataclysm that followed:

> Oliphant came over from England and attended a meeting of the Uranium Committee … He realised that something was very wrong and that the work on uranium was not being pushed in an effective way.
> He travelled across this continent from the Atlantic to the Pacific and disregarding international etiquette told all those who were willing to listen what he thought of us. Considerations other than those of military security prevent me from revealing the exact expressions which he used. If Congress knew the true history of the atomic energy project, it would create a special medal to be given to meddling foreigners for distinguished services, and Dr Oliphant would be the first to receive one.[15]

* * *

Oliphant's intention to bend the will of a reluctant nation could never have been accomplished without an equally insistent American co-conspirator. Oliphant acknowledged this in the letter he sent to Lawrence from Washington before his flight back across the Atlantic.

'May I say how much I enjoyed my few hours in Berkeley, and how much I still admire the way in which things are done in your Laboratory,' he wrote. 'I feel quite sure that in your hands the uranium question will receive proper and complete attention, and I do hope that you are able to do something in the matter.'[16]

Oliphant returned to Birmingham, satisfied that his efforts would prove worthwhile, but scarcely imagining how rapidly events would move. 'Ernest took this up like anything,' he would remember of Lawrence's subsequent actions. 'He was a very vigorous person. Before you could say Jack Robinson, he'd moved a committee, and then Bush and Conant, and finally the President to set up the Manhattan Project.'[17]

That cascade of events was initiated by Lawrence's call to Arthur Compton, which brought an unexpected outcome. Compton agreed that Lawrence should make his case directly to Conant, who, as chance would have it, was due to be at the University of Chicago the following week as part of that institution's fiftieth anniversary celebrations. On the same evening, Lawrence would be delivering a lecture to Chicago's Museum of Science and Industry, on the potential medical applications spawned by his new cyclotron.

The three men met later that night at Compton's house, taking coffee around a raging fire as the autumn chill set in. Lawrence presented his own précis of Oliphant's MAUD summary, and once again bemoaned the lack of action coming from Washington despite the growing fear – fanned by Oliphant – that Germany was moving forward with its atomic bomb plans.

It was then that Conant, who did not reveal he was fully aware of the MAUD Report and its findings, pulled his own surprise. He admonished Lawrence for his egregious lapse of national security in revealing unauthorised details to Oppenheimer about bomb research.

If Lawrence felt he was taking the fall for Oliphant's indiscretion at Berkeley, his irritation might have been mollified had he known Conant had expressed similar misgivings about the Australian's trustworthiness to Bush a week earlier: 'Oliphant's behaviour does not help the cause of secrecy!' Conant had sniped.[18]

Conant then tested Lawrence's commitment to the cause by suggesting the services of America's physicists could be better utilised in areas of defence work other than an unproven atomic bomb concept. For final effect, he added that he had advised Bush to terminate the fission investigation, and redeploy those on the uranium committee in more viable programs, such as radar.

When Lawrence took the bait and began pointing out the folly of such a move, Conant reeled him in.

'Ernest, you say you are convinced of the importance of these fission bombs,' Conant said, his gaze shifting from the fireplace to the physicist. 'Are you willing to devote the next several years of your life to getting them made?'

Lawrence's jaw dropped and his mouth gaped wide as the gravity of the request, and the position in which it placed him, became clear.

In another of those history-defining moments that littered those few September weeks, Lawrence sifted his thoughts with sufficient clarity to respond: 'If you tell me this is my job, I'll do it.'[19]

From that evening, Lawrence became as ardent in his support for the atomic bomb as Oliphant had been during his campaigning across the United States. When the uranium committee next met in October 1941, Lawrence opened the meeting by reading aloud Oliphant's summary of the MAUD Report.

Within two weeks of Oliphant's return to Britain, Vannevar Bush took a copy of the MAUD Report to the White House and presented its findings, as well as a page of talking points drawn up by Conant, directly to President Roosevelt. So convincing was Bush's case that Roosevelt agreed to the immediate launch of a comprehensive research program with the stated outcome of building a uranium bomb.

On 11 October 1941, less than three weeks after Oliphant and Lawrence had stood atop Charter Hill and fused their formidable wills, Roosevelt penned a letter to Churchill to set in motion the process that would yield the most destructive force humankind had ever imagined.

'My dear Winston,' wrote the president of the nation that, months earlier, had decided on the basis of two top-level reviews that a bomb was not an immediate priority, 'It appears desirable that we should soon correspond or converse concerning the subject which is under study by your MAUD committee, and by Dr Bush's organisation in this country, in order that any extended efforts may be co-ordinated or even jointly conducted.'[20]

Years before the bomb project had yielded a weapon, it hosted a pre-emptive chain reaction that altered history's course. The energy brought across the Atlantic by Mark Oliphant had been infused into Lawrence, who had projected it to Conant, then Bush and finally to the Oval Office.

* * *

As events played out, the timing proved as momentous as the decision itself. Barely two months after Roosevelt's missive to Churchill, Japan launched a bombing raid on the United States Pacific Fleet stationed at Pearl Harbor.

Had the ink not been barely dry on final approvals for the bomb project, it was unlikely to have gained top presidential priority in such a radically altered political climate. It's also not unreasonable to assume, therefore, that the Second World War would have passed without the delivery of the atomic bomb's horrifying exclamation mark.

The furious entry of Japan into the war also changed life for Mark Oliphant. Having returned from the United States to be confronted by the disbandment of the MAUD Committee, Oliphant eventually settled back into work at Birmingham. He retained his place on the Tube Alloys project's technical committee, although his input was minimal due to the ongoing demands of Birmingham's radar work, where new and improved units continued to be produced in the laboratory.

However, the impact of his unofficial American sojourn had been noticed in higher offices.

Birmingham Vice-Chancellor Raymond Priestley wrote in his diary in early October 1941: 'Oliphant is back from America – by bomber – and Charles [Wright], who came down to see the lab last week and stayed the night with us, says that he has gone down very well out there. He is so impressed with the value of Oliphant's work, indeed, that he would like to see him have some reward.'[21] Wright, director of scientific research with the Admiralty, in addition to being Priestley's brother-in-law, was sufficiently well placed to hear whispers from power's corridors.

Even if Oliphant had been aware of any such talk, he doubtless would have dismissed it as mere background noise. He was, despite his stature within the global physics community and the decisive advantages he had helped deliver to the Allied war effort, driven by his dedication to science and not any personal kudos it might bring. Besides, barely had he returned from his mission to America than his attention was drawn to other events that were moving at pace even further afield.

20

A MISGUIDED MISSION

To Australia and back, 1942 to 1943

Mercifully removed from Britain's wartime torment she might have been, but life had not been correspondingly straightforward for Rosa Oliphant. Any comfort she might have wished for upon return to the home shores her two young children had not previously seen diminished just weeks before they sailed. On top of leaving behind her husband and her home, Rosa departed England having learned that her mother, Clara, had died suddenly in hospital aged seventy-eight. (Her father, Frederick, had died many years earlier.)

To compound the hardship of her relocated life as a single mum, Rosa's hopes of finding work to supplement her £20 monthly allowance from Mark's salary (around AU$1700 today) were stymied, as in Cambridge, by her lack of previous employment. 'I wanted to get work, but I was not trained to do anything. [There was] no kindergarten, no pre-school, no child-minding facilities.'[1]

She was understandably jubilant, therefore, when a telegram arrived from her husband in February 1942, stating simply 'am being sent to Australia'. The reason for this unforeseen occurrence would soon become clear.

* * *

When Singapore fell to the Japanese in mid-February 1942, the threat of imminent invasion that had compelled Oliphant to repatriate his wife and children from Britain was thought to be levelled at Australia. While concern for his family was a primary motivation in his desire to travel to the homeland he had not seen in more than a decade, his rationale also reflected the call of patriotic duty.

Oliphant's restive character, enlivened by the impact of his evangelical visit to the United States, demanded a new challenge. The violent threat posed by Japanese aggression towards his Australian homeland was just the scenario that might benefit from his expertise.

True to type, he was not about to sit on his hands awaiting an invitation. No sooner was Singapore's fate decided than Oliphant contacted Australia's high commissioner in London – former prime minister Stanley Bruce – offering his services to the nation's incumbent leader, John Curtin. He followed this up with a cable to the University of Sydney's John Madsen, who was Australia's kingpin in radar research at the time. His missive was passed on to the head of the Council of Scientific and Industrial Research in Melbourne, Sir David Rivett.

Oliphant was already known to Rivett, through an earlier indiscretion that had arisen from his loose-lipped American campaign. Shortly before flying west to meet with Lawrence in September 1941, Oliphant had shared dinner with Australia's inaugural United States ambassador in Washington – another former federal MP, Richard Casey – who was surprised to hear his new acquaintance revealing details of the highly secret uranium research being undertaken in Britain. His curiosity piqued, Casey asked Oliphant to provide a brief written summary of the MAUD Committee's findings relating to atomic energy (though not details of any proposed bomb), which was duly delivered to the embassy next day. It was the first formal notification Australia had received of prospects for nuclear power derived from uranium fission.

Casey wasted no time in forwarding Oliphant's report to Rivett, with the accompanying note that Australia might want to get involved in any developments for the 'energy machine'[2] before all its patents were locked away by other administrations. Not only had Oliphant wilfully bypassed security provisions in discussing the MAUD Report with unauthorised

American colleagues, he had also chosen to admit a previously uninvited player into the security bubble.

It was therefore with a growing reputation for indiscretion that Oliphant announced he wished to take leave from Birmingham.

'Oliphant came to supper … and told me that he might have to go to Australia to take charge of radiolocation there,' the university's vice-chancellor Raymond Priestley wrote in late February. 'If they want him we shall have to let him go, for Australia, with its wide spaces and sparse population, is particularly dependent upon radio of all sorts for its defence.'[3]

By that time, the first airborne raid on Darwin had already been launched.

* * *

When he received the standard cryptic message in mid-March to travel via train to Scotland, Oliphant prepared himself for a gruelling, even longer journey aboard a bomber aircraft to Australia. He was surprised, and more than a little miffed, when follow-up orders directed him instead to Glasgow docks, where passage had been reserved on a commissioned troopship that had previously plied the Antipodean route as the *Dominion Monarch*.

The prospect of indolent weeks at sea rather than a few days of discomfort in the air set the impatient physicist immediately on edge. While the risk of such a long voyage through treacherous wartime shipping channels was mitigated by the convoy of almost fifty vessels that accompanied the *Dominion Monarch* – including half a dozen destroyers, a pocket battleship and the 23,000-ton aircraft carrier HMS *Illustrious* – Oliphant soon sensed he had committed a serious tactical error. As the fleet weighed anchor and nosed into the North Atlantic, beneath which the torpedoed *Athenia* now lay silent, he began to wonder whether he would have been of greater service back in his office and laboratory at Birmingham.

After a month at sea, during which he passed his time presenting lectures to interested crew and officers, Oliphant arrived in Cape Town with a further six weeks' sailing ahead of him. Despite advising Rosa he was on his way shortly before leaving Glasgow, Oliphant used his African stopover to send an urgent cable to Stanley Bruce in London, requesting that room be found for him on the next available transport back to Britain. He made little effort to hide his grievance at the High Commission's failure to secure him air travel.

Bruce's response was curt. Wartime transport was a fraught logistical exercise that did not accommodate the whims and fancies of the only non-military passenger on a troopship. By the time that word of Oliphant's petulant change of plans reached Australia and it was agreed that radar development there could continue without him if circumstances demanded, he was back aboard the liner codenamed SL-4 – heading eastwards across the vast Indian Ocean.

* * *

As the weeks passed, Rosa Oliphant's anxiety grew. She had heard nothing from, or of, her husband since his cable from Glasgow. Each day without news rendered her privately more fretful. She even wrote to Rivett, who erroneously advised – based on his knowledge of Oliphant's communications from Cape Town – that he believed her husband was in the process of returning to England.

Then, on the evening of 27 May, ten weeks after Mark set off, Rosa's telephone rang and he crackled down a faltering line from Fremantle, where he had made safe landfall. He advised that next morning he would board a plane for the 2700-kilometre onward leg to Adelaide, before the call cut out. It had been almost two years since she had heard that voice.

Rosa's relief paled alongside six-year-old Michael's unrestrained glee. He greeted his mother's revelation with his own announcement: 'when I see daddy, I'll run and run'.[4] Next morning, Rosa rang Baron and Beatrice Oliphant to share the news, and the entire family – including four-year-old Vivian – drove to Parafield Airport, beyond Adelaide's northern fringe, to await Mark's homecoming.

The DC-3's whirring twin propellers shut down and it taxied to a standstill at the apron of the windswept tarmac, where a cold south-westerly howled to herald the imminent arrival of winter. As the familiar frame of Mark Oliphant eased itself down the rear stairs and onto home soil, none of the waiting welcome party could quite believe their eyes.

In the years since his family had last seen saw him, his woolly shock of auburn-brown hair had turned totally white – a fact he had neglected to share in his regular written communications. Baron subsequently told the family he himself had undergone a similar transformation early in middle age, but as Mark walked towards the group it was clear that the stress of work and war had taken their toll on the forty-one-year-old professor.

Once reassured that the obvious change did not mean his father was suddenly ancient, Michael made good his earlier promise and raced towards him, before being scooped up in his father's arms and clinging tightly to his neck.

The young boy's euphoria would be short-lived. When Rosa suggested they squeeze back into the car and return home for a proper reunion, Mark looked at her quizzically.

'Oh no,' he said, apparently unaware that full details of his schedule had not been shared. 'I have to go on the plane to Melbourne.'

Once the DC-3 was refuelled and the children told that their father, two years absent, would not be coming home with them after all, Mark climbed back aboard and disappeared into the scudding clouds.

As stunned and disappointed as she was, Rosa's immediate empathy lay elsewhere. 'His poor parents,' she would recall with the stoicism demanded of wartime wives. 'He was on duty.'[5]

* * *

After Melbourne, where he convened briefly with Rivett, Oliphant was again on the move. An overnight train took him 900 kilometres to Sydney and a meeting with the radar research division of the Council for Scientific and Industrial Research. It became immediately obvious that, during the course of his voyage from Britain, plans had moved on and he was essentially surplus to requirements.

It was not the only development he had missed while captive aboard the SL-4.

Days out from his disembarkation at Fremantle, a letter addressed to Oliphant at Birmingham University arrived from the British prime minister's private secretary at 10 Downing Street. If Oliphant's weakness for revealing secrets was causing heartburn within the security and science communities, it had clearly not been relayed to the top office. As Charles Wright had foreshadowed, Oliphant was indeed being considered for 'some reward'. Winston Churchill had agreed that Oliphant's name should be submitted to George VI for inclusion in the forthcoming King's Birthday Honours List for appointment as Officer of the Order of the British Empire.

The problem for colleagues of Mark Oliphant, prospective OBE, was that none of them quite knew where he was. Or how to alert him.

Robert Nimmo, one of the unfortunate research assistants to have had his legs broken in the cyclotron's construction mishap, stalled for time with Downing Street, which required confirmation that the honour would be accepted. Nimmo fired off an urgent cable to Rivett's office in Melbourne, pending Oliphant's expected arrival there. 'Your name to be submitted Birthday Honours Officer Order British Empire if you agree. Awaiting your reply.'[6]

A week passed before Nimmo received word from Oliphant, in a terse telegram that reflected his state of mind after meetings in Sydney: 'Honours belong laboratory so please refuse my behalf. Position here very unsatisfactory. Regards Oliphant.'[7]

That response was then reframed in suitable diplomat-speak and Churchill's office advised that, while Oliphant was deeply appreciative of the recognition afforded by the Prime Minister's suggestion, he believed any success his work had brought was a team triumph, and deserved to be rewarded as such.

The next day, Churchill's secretary responded: 'The Prime Minister is very sorry not to be able to include the Professor's name in his list of recommendations to the King but he will, of course, respect Professor Oliphant's wishes and proceed no further with his proposal.'[8]

* * *

For the next two months, Oliphant's annoyance simmered. Suggestions that it had been the input of radar research supremo John Madsen that saw him deprived of air transport to Australia fed his discontent. And not even the acclaim with which he was greeted when he was invited to visit Rutherford's homeland on radar business ameliorated the view that he had wasted much time and effort. In more celebratory times, a sojourn in New Zealand might have afforded Oliphant a chance to see the country and gain insight into the experiences that had shaped his mentor's life. But Oliphant's focus remained strictly business; his fleeting trip took him directly to Wellington, where he addressed a meeting of New Zealand scientists engaged in radar work, and then straight back to Australia.

So desperate had Oliphant become to make up for the lost months that he secured passage back to Britain for himself, Rosa and the children – whom he had seen only on occasional visits during his stay in Australia – upon the *Desirade*. The vessel was operated by Free French interests, but crewed by a hybrid of Vichy Government supporters and those loyal to Charles de Gaulle's alternative administration. Conditions aboard the vessel were equally chaotic, with the Oliphants offered no fresh towels or clean bed linen throughout the weeks it took to reach Durban on South Africa's east coast. Naval authorities there promptly deemed the *Desirade* unfit for onward passenger travel.

While the family took accommodation in South Africa's inland hills, where they celebrated Christmas 1942, Oliphant wired another plaintive request to his Admiralty contacts for urgent alternative transport. However, it took a further month for that to happen, and it was not until the final day of northern winter in February 1943 that the family landed at Glasgow, then headed to their beloved Peto at Barnt Green.

Almost eleven months had elapsed since Mark Oliphant had left Britain on his misguided patriotic mission, of which almost half he had spent in transit.

* * *

Morale in Britain had shifted notably in his absence. The German threat remained, but vital victories had been won in North Africa and the airborne radar Oliphant had helped beget was guiding Allied bombing raids on targets across Western Europe. The air-raid shelter that Oliphant had built at Peto at the height of the Blitz was now appropriated by Michael and Vivian as a playhouse. And the only visible hints of imminent peril in rural Worcestershire were the occasional barrage balloons that still appeared above the Lickey Hills' treeline during the family's weekend walks.

If a benefit other than reunion had emerged from this trying chapter, it was that Oliphant's enforced confinement at sea had – in the great Rutherfordian tradition – afforded him much time to think. It also meant that when he re-entered his Birmingham laboratory, which had continued to be consumed by radar work throughout his absence, he was in a characteristic rush to get things moving.

Due to his voluntary relocation, he had lost touch with work being carried out under the Tube Alloys program. But once back home, it didn't take long for him to establish that progress on the uranium issue in Britain had been minimal. He believed a key reason was that the gaseous diffusion method of separating the uranium-235 isotope recommended in the MAUD Report was too technically cumbersome. Oliphant now believed the method he had discussed with Lawrence under the shadow of the giant cyclotron at Berkeley eighteen months earlier was more feasible.

Within weeks of returning to Birmingham, where he felt the radar program had served its purpose and 'shot its bolt', Oliphant informed vice-chancellor Raymond Priestley as well as Tizard that he wanted his laboratory team and resources to be directed to nuclear physics and, in particular, how that might help with the bomb project he had so passionately championed. While it seemed an abrupt decision, it was another that he had reached during his lengthy confinement at sea.

On 26 May 1943 he penned a confidential letter to Sir Edward Appleton, upon whom he had unloaded his frustrations at the establishment of the Tube Alloys project eighteen months earlier, suggesting this new large-scale means of isotope separation 'might remove the [uranium] project from the realms of gigantic chemical engineering and render it a more practical proposition in this country'.[9]

Oliphant understood that the bomb program had become essentially an American enterprise, from which British expertise had been conspicuously excluded. However, his need to be at the heart of the biggest issues as they related to his field of inquiry, coupled with his belief that the progress of science should not be constrained by cultural or geographic boundaries, meant he pursued his new quest with trademark vigour.

His vision was to employ electromagnetic separation techniques to isolate quantities of the desired but elusive uranium-235. Firing charged uranium atoms into a magnetic field that bent them at high speed around a circular track meant that those with the heavier atomic mass (uranium-238) would gravitate to the outside, while the lighter 235 particles would travel in a beam closer to the centre, where they could be harvested.

The hope was that the still-unfinished sixty-inch cyclotron at Birmingham could be reconfigured to conduct this work. According to Oliphant's upbeat estimates, if sufficient power were applied to his giant magnet then the apparatus could yield around a kilogram of enriched uranium within a couple of weeks. There would then be enough to form the required critical mass for an atomic bomb inside a few months, and that would surely reinvigorate the nuclear program in Britain.

While the legitimacy of these numbers raised eyebrows among some, the premise was deemed worthy of further investigation by Tube Alloys' hierarchy.

Yet Oliphant's hasty decision to terminate his laboratory's involvement in radar and direct those resources to the question of nuclear fission was not warmly welcomed by Henry Tizard, or by Birmingham's vice-chancellor. Having seen his physics professor take a year's sabbatical to pursue an unnecessary (as it transpired) cause in Australia, Raymond Priestley noted sourly that Oliphant had promptly 'ratted' from the radar work in which Birmingham had been heavily involved to the Tube Alloys program soon after his return.[10]

Part of Priestley's disquiet stemmed from the fact that Oliphant's laboratory had been fitted out and funded by the Admiralty, to enable the upscaling of the radar research program. Not only had that delivered valuable financial benefits to the institution, but it had also earned Birmingham a measure of wartime prestige.

However, once Oliphant made it clear that there was little more he could add to the already substantial capabilities of radar, and that the war effort might benefit more from his uranium endeavours, the differences were sorted out. Despite having been such a strident critic of the Tube Alloys structure when it had first been unveiled in late 1941, less than two years later he became part of its machinery.

Not that his past imprudence was completely glossed over. In June 1943, Wallace Akers – whose Tube Alloys appointment Oliphant had previously savaged as 'disgraceful' – wrote to Appleton to query the specifics of their agreement with the Australian, whom Akers suspected might still be tempted to split his time between nuclear and radar research.

> I must say that we were most definite that Oliphant whole-time would be a most valuable addition to our effort but if he proposes to work part-time on this we would certainly not want him. Such an arrangement would be ineffectual and dangerous. I say the latter because Oliphant is, as you know, impetuous and none too discreet, so we would not want to let him in on all the secrets of TA work unless he is properly tied to us.[11]

But before Oliphant could set up his cyclotron to make meaningful headway on electromagnetic separation of uranium, politics once again intervened.

* * *

In the almost two years that had passed since Oliphant had toured the United States, beseeching all who would listen to pour their best efforts into realising a bomb, the power balance between the two nations had swung drastically. The leadership Britain had shown on the nuclear question had dissipated, despite Roosevelt's suggestion of a 'co-ordinated or even jointly conducted' approach, while America's commitment to an atomic weapon had forged forward at pace.

Crucially, it was Roosevelt's decision to entrust the project to the United States Army Corps of Engineers, rather than a privately led enterprise like Tube Alloys, that had brought such decisive progress. By August 1942, the corps had established a dedicated engineers' precinct to deliver the president's aim. While that jurisdiction would eventually spread out to a number of secret locations across the United States, its heart was an office building on Broadway in New York City's downtown Manhattan. As a result, the entire operation was given the nondescript code name 'Manhattan Engineer District', even though it transcended all geographical boundaries, before it became known simply as the 'Manhattan Project'.[12]

The ground that Oliphant had so assiduously tilled in the hope that America might help sow it had, in the space of barely a year, been summarily claimed by Britain's ally. And by late 1942, when it had become obvious that the United States had both the will and the manufacturing capability to complete the project they had for so long regarded with scepticism, Britain realised it was in danger of being excluded.

The demarcation became more pronounced when the highly abrasive but hugely effective General Leslie Groves – previously in charge of construction for the entire United States Army, and fresh from completion of the Pentagon – was installed as the Manhattan Project's overlord.

The impasse was formally breached in August 1943, when Churchill and Roosevelt met secretly in Quebec and agreed to a shared commitment to building nuclear weapons for the duration of the war. What it meant for post-war weapons development would only become public in the conflict's aftermath.

Such was the hurry to insert themselves into the Manhattan Project that the first British delegation – led by James Chadwick, with Oliphant and Rudolf Peierls – was booked to fly to New York before the Quebec Agreement had been formally signed. It was an eagerness interpreted by some in the American science and military communities as representing 'almost indecent haste'.[13]

Certainly that suspicion must also have weighed upon Rosa Oliphant, who had only just resettled into British life in 1943 and enrolled the children into new schools. Military protocols did not extend to transport and resettlement costs for next of kin, so they would remain at Barnt Green while Oliphant prepared for another absence of indeterminate length – and for reasons unexplained.

For Mark Oliphant, however, another great adventure suddenly loomed.

21

MANHATTAN

United States, 1943 to 1945

Rather than being airlifted by bomber, Oliphant and his colleagues would take the flying boat service from Foynes, on the banks of western Ireland's Shannon Estuary. With their New York bound plane delayed, the group passed time at a cinema in nearby Limerick, where the feature film was preceded by a newsreel that included predictions of an atomic bomb made possible by uranium chain reactions.

The delegation's initial forays into the United States were hardly spectacular. On their first day in Washington, Oliphant and Peierls were deep in discussion and crossed a street against a pedestrian light, which was noticed by a vigilant traffic policeman. The pair were only spared a fine after producing travel documents that proved their new-arrival status.

Chadwick – who, under the bilateral structure, had effectively replaced Akers as head of Britain's nuclear research – was not so fortunate. Having not previously visited the United States, he received pre-departure eating recommendations from Lady Astor, the American-born socialite who became the first woman elected to British Parliament. Chadwick was told not to miss Harvey's Oyster Bar in New York.

'Soon after arriving, we took a taxi to the recommended restaurant where Chadwick ordered, and ate, a dozen oysters,' recalled Oliphant, who was also sharing a hotel room with his former Cavendish colleague.

'Later that night he became terribly ill, and in the morning was still obviously suffering. But he insisted we fly to Washington ... I was very anxious as I saw an extremely sick but determined man into his car.'[1]

* * *

When Oliphant arrived with his colleagues at the Pentagon – which looked and smelled as pristine as the Quebec Agreement, now executed and delivered to the White House – he was pleased to be greeted by the familiar figure of Robert Oppenheimer. Less welcoming was the foreboding presence of General Groves, whose frightening reputation was well known within Washington circles and far beyond.

Son of a Presbyterian minister, Groves ran projects with a singularity that bordered on despotism. One of his key deputies on the bomb project, Colonel Kenneth Nichols, described him as 'the biggest S.O.B. [son of a bitch] I have ever worked for. He is most demanding. He is most critical. He is always a driver, never a praiser. He is abrasive and sarcastic. He disregards all normal organisational channels ... He is the most egotistical man I know. He knows he is right and so sticks by his decision.'[2]

Upon gaining responsibility for the Manhattan Project in September 1942, Groves began his first meeting with a group of the operation's American-based scientific elite that included Oppenheimer and Arthur Compton by growling 'you're working for me now and you'd better toe the line'.[3] He saw the members of the physics fraternity as military conscripts, albeit out of uniform.

Groves was also unrelenting in pursuit of the Manhattan Project's grandiose ambitions. It would be the largest-ever commitment of personnel and resources to realising a single scientific quest. Over the course of three years, it would employ more than 200,000 people and absorb an estimated US$2 billion (approximately US$25 billion today). And none of those who attended that meeting at the Pentagon on 10 September 1943 left it with any doubt as to the result they were chasing.

'It was emphasized again and again,' Oliphant would remember vividly, 'that the prime object was the production of a military weapon in the shortest possible time.'[4]

Groves disliked and distrusted the British, both personally and professionally, a resentment born of his experiences in the First World War. He would later excoriate 'the general air of superiority which seems to be inherent in the English. Only the English could be so annoying towards everyone not of their nationality.'[5] Groves was also suspicious of the commercial motives behind ICI's involvement in Tube Alloys in Britain, and took pride in restricting interactions between individuals on either side of the Atlantic. In particular, he was deeply distrustful of Wallace Akers, who remained in charge of a depleted Tube Alloys in England.

Groves held clear ideas on how the British contingent might best be absorbed into the Manhattan Project, if they must be involved at all. He wanted Chadwick and Oliphant to head directly to Oppenheimer's Los Alamos laboratory, perched atop steep cliffs overlooking the New Mexico desert. It was one of more than a dozen project sites spread across the United States, all designed to operate in strict isolation and with no interchange of information except at the highest level, as per Groves' inviolable instruction.

Over the next three years, the Manhattan Project would reach across the length and breadth of America, and beyond. Research into uranium separation and fission experiments was conducted at numerous universities, including Lawrence's Berkeley, Columbia in New York, and the University of Chicago, where the first controlled nuclear fission reaction was successfully achieved in December 1942. That historic moment, when the world's first nuclear reactor – constructed on a squash court beneath the grandstands of Chicago University's Stagg Field – underwent a self-sustaining chain reaction, used more than five tons of naturally occurring uranium metal and was overseen by Italian-born physicist Enrico Fermi.

A huge uranium enrichment facility was established at a greenfield site in Tennessee, which later became known as the township of Oak Ridge. In addition to the heavy water production site located alongside the Chalk River in Ontario, Canada, the world's first large-scale nuclear reactor would be built south of the Canadian border, at Hanford in Washington State. And the heart of bomb design and production would be the highly secret site at Los Alamos.

Groves had hand-picked Oppenheimer to lead the weapons development operation – a controversial choice to many, given Oppenheimer's known connections with Communist activities and sympathisers. Groves himself had initially been unconvinced of Oppenheimer's suitability, because he boasted neither administrative experience nor a Nobel Prize.

It was Oppenheimer who had selected the laboratory's top-secret location. As a teenager, he had contracted a near-fatal bout of dysentery and spent a summer convalescing at a dude ranch in the Sangre de Cristo Mountains near the New Mexico capital Santa Fe. He saw this as an ideal site where such an audaciously secretive community might hide. The clandestine outpost would become home to thousands of workers, yet was designated by a single registered post-office box, and was known within the Manhattan Project's inner sanctum simply as 'Site Y'.

After his unpleasant experiences at the Cavendish Laboratory, Oppenheimer had established himself as a formidable theoretical physicist. Before arriving at Berkeley, he had studied alongside Max Born and Niels Bohr at Germany's Göttingen University in the mid-1920s, at the time they were developing the theory of quantum mechanics. A tall, aloof presence, Oppenheimer had a passion for philosophy and Eastern religion, for English and French literature, for fine arts and dry martinis that rendered him the antithesis of everything Groves represented. Yet the physicist's brooding charisma made him hugely successful in securing recruits to work for him, even though many of them had no clue as to the project's ultimate goal.

From the outset, however, Oliphant did not see Los Alamos as the detail that best suited him. His interest was in the science of isotope extraction, not the complex ordnance calculations on how the bomb might be constructed, which occupied the great theoretical minds gathered in the New Mexico desert. Oliphant also saw far more appeal in working alongside his great friend Ernest Lawrence at the Berkeley lab that had hosted research into electromagnetic separation techniques for the past two years.

Groves voiced surprise at Oliphant's reluctance to follow Oppenheimer – who also wanted to work with the Australian – back to the isolated compound atop the red-earth mesa of New Mexico's Pajarito

Plateau. It would not be the last time that Oliphant would challenge Groves' iron will.

On this occasion, though, he was supported in his commitment to Berkeley by Chadwick, and in the end Groves recognised pragmatism ahead of his prejudices. So in early November 1943, Oliphant hauled himself into the depths of another B-24 bomber to take up his tenure as assistant – effectively deputy – to Ernest Lawrence on the shores of San Francisco Bay.

* * *

The first item that caught Oliphant's attention on arrival at Lawrence's California laboratory was the huge 184-inch cyclotron that had been taking shape on his previous visit, but was now completed and housed within a striking twenty-four-sided structure. The next was the sheer scale of the operation in which he had become embedded.

In the wake of Oliphant's 1941 crusade, Lawrence's faith in the capacity of his cyclotron to successfully separate the required uranium-235 isotope had grown at the same rapid rate as the scope of his machines. While the practicality of the electromagnetic separation method was doubted by some physicists, Lawrence threw himself into the task of finding ways to improve its yields. As the cyclotron evolved, so too did its name change to the 'calutron', which recognised its conception in the radiation laboratory at Lawrence's California University.

Among the changes Lawrence had introduced was the use of 'hot' (high-positive-voltage) electrical energy, as opposed to a 'cold' (grounded) power source. Hot energy reduced the risk that insulators would fail, and therefore allowed the machine to make more efficient use of electricity. It also meant the calutron could employ multiple streams of charged particles rather than a single beam, which bolstered the potential for production of the fissionable isotope. As these charged uranium particles were fired through a strong magnetic field, the lighter and sought-after uranium-235 isotopes would be deflected more distinctly than the heavier uranium-238, which meant they could be collected in greater volumes (although still atom by atom) in the machine's dedicated receivers.

Once tested and further streamlined, this powerful machine would be replicated many times over on the bare expanse of Oak Ridge – later known as 'Site X' – where, from late 1943, the 'racetracks' around which the precious particles surged began slowly churning out the material needed to form the core of a bomb. The coils of the magnets these machines used were wound with silver sourced from the Fort Knox bullion depository, a total of about 14,000 tons that cost around US$500 million.

While the weapons work was under way at Los Alamos, Oak Ridge was where the infrastructure for the different isotope separation methods being explored would be housed. Among Oliphant's first assignments after arriving at Berkeley was to visit this facility, where he was stunned by the amount of building that had already taken place. Not even the all-pervasive layer of sticky, ochre mud that coagulated under late autumn rain on the Tennessee Valley floor could slow the pace of construction, which yielded up to 1000 new homes per month. Oak Ridge would eventually grow to house more than 50,000 workers, with the true nature of their task unknown to anyone outside the Manhattan Project and the compound's regularly patrolled barbed-wire fences.

The Oak Ridge facility ultimately hosted the three methods of isotope extraction that Groves and his scientists believed would most reliably produce sufficient uranium-235 for an atomic bomb. Under Groves' relentless leadership, the three refinement methods – gaseous diffusion, electromagnetic separation and, eventually, thermal diffusion, which Otto Frisch had originally pursued at Birmingham – were effectively pitted against one another in a contest to prove which was the most productive.

Although the thermal diffusion technique had been largely dismissed as being capable of yielding significant quantities of uranium-235, it was incorporated into the Oak Ridge set-up (in a plant codenamed S-50) in late 1944. By that stage, with the race to build the bomb becoming more and more urgent, all three separation methods were contributing to the precious uranium-235 resource.

Initially, it had been expected that a majority of the enriched material would be collected through the gaseous diffusion process that had been recommended in the MAUD Report. However, the fears Oliphant expressed upon returning to Britain in early 1943 that the gaseous process was too cumbersome were soon revealed to be prudent. The intricate network of pumps and porous barriers that were needed to treat the toxic uranium hexafluoride gas in order to remove the rare uranium-235 isotope was installed in a huge plant at Oak Ridge codenamed K-25. It represented an extraordinarily ambitious engineering feat, which was regularly beset by breakdowns and maintenance problems.

Oliphant's mistrust of the gaseous diffusion method only grew after he had spent some time at Oak Ridge. It was scarcely surprising that he identified shortcomings, given the nature of the operation, which took laboratory experiments still effectively in their infancy and applied them on a mega-industrial scale in a race against time. The gaseous diffusion process relied upon a series of 4000 stages, each one dependent on the success of the other, thereby making it highly susceptible to shutdown should one of those steps malfunction. It also required the uranium hexafluoride gas to be pumped through an endless network of atomic sieves set with holes four-ten millionths of an inch (ten nanometres) in diameter.

As a consequence, Oliphant's support swung further behind electromagnetic separation, which employed a series of Lawrence's calutrons arranged side by side in the huge elliptical 'racetracks'. At the height of isotope production in late 1944, the electromagnetic facility, known as 'Y-12', housed nine racetracks comprised of more than 850 calutrons.

In contrast to the temperamental and inter-related technology of the gaseous diffusion plant, the electromagnetic technique consisted of a huge number of standalone units that could operate in parallel rather than relying on the performance of each individual component. Not that it quarantined the fledgling technical operation from serious setbacks, as Oliphant quickly discovered. When he paid his first visit to Oak Ridge in late 1943, the calutron racetracks had ground to a complete standstill after rust and other impurities were found in the oil that circulated to keep the huge magnets cool. The repair job took months, and required the magnets to be taken apart, meticulously cleaned and then rewound.

One reason why Lawrence pressed strongly within the Manhattan Project for Oliphant to join him in the separation work was that his Australian friend's insight and experience – gained from working alongside Rutherford and then designing his own (still-unfinished) cyclotron at Birmingham – provided a natural fit with the application of electromagnetism. And Oliphant duly hurled himself into his role as Lawrence's right-hand man with customary vigour.

Much of the expertise that Oliphant was to provide stemmed from the skill that had first won him notice as a scientist: his ability to troubleshoot equipment problems. Since it had started to take shape in mid-1943, the Y-12 electromagnetic plant had been plagued by faults that arose from its novelty and complexity. The calutrons, which stood vertically, initially suffered from vacuum tank leaks, often because they were knocked out of alignment by the massive magnetic pressure being generated. Welds regularly failed, electrical circuits shorted out, and it took time for the machines' operators to learn the tricks and pitfalls of handling such temperamental apparatus.

The work that Oliphant had previously envisaged undertaking at his cyclotron in Birmingham was being pursued in a different manner by Lawrence, through the use of his calutron, which (unlike the cyclotron from which it had evolved) relied on magnetic force and did not utilise large-scale electric fields. But as Oliphant familiarised himself with the equipment and its methods, his faith in this new technology began to mirror Lawrence's.

In November 1943, just months after joining the Manhattan Project, Oliphant wrote to General Groves and betrayed some of his own frustration that the ongoing teething problems delivered. 'The electromagnetic method as developed [at Berkeley], and as under construction at Site X [Oak Ridge], is without doubt capable of the performance claimed for it,' he wrote. 'Beyond the accidents and failures due to faulty mechanical and electrical design, which are difficult to condone, the electromagnetic separation system is free from all uncertainties.'[6]

In Oliphant's assessment, once the electromagnetic method's operational issues were smoothed out, a critical mass of uranium-235 – perhaps 20–30 kilograms – should be successfully harvested by the end of the following year, 1944. This would happen even sooner if Groves took up Oliphant's pointed suggestion that the size of the electromagnetic plant be increased 'at once by a factor of five'.

'In fact, I would go further to say that unless this is done I cannot see that the nuclear weapon will be of military value in this war,' Oliphant bluntly warned.[7]

It was one of a number of recommendations from Oliphant that the general chose to ignore.

* * *

Oliphant's role as Lawrence's trusted deputy meant he spent his time oscillating between Oak Ridge, Washington and Berkeley (the environment he preferred), and even made one fleeting visit to Los Alamos. But he was also effectively second-in-charge to Chadwick among the British delegation. As a result, his counsel was often sought by his former Cavendish colleague, even though Chadwick was the only member of the team with authority to know what was happening across other sectors of the Manhattan Project.

It was in the guise of notional second-in-command that Oliphant would make numerous trips between the United States and Britain over the remainder of the war. They involved a sixteen-hour journey that proved a trial for even the hardiest of flyers, of which Chadwick was not one, which meant much of the essential commuting was delegated to Oliphant.

Conditions were far more rudimentary than the flight-deck privileges he had enjoyed aboard the B-24 from Scotland to start his lobbying mission two years earlier. As merely another body passing through the revolving door of academics, bureaucrats and military brass carried over the Atlantic, he was relegated to the bomb bay along with other rank-and-file travellers.

Rather than a jockey seat in the cockpit, he was routinely allocated a thin mattress spread over the aircraft's bomb doors. He was able to manoeuvre his lanky frame, clad in a bulky flying suit, to jam his boots against one side of the metal fuselage and push his head against the edge opposite, using his parachute pack as a pillow. The constant roar of four super-charged engines, coupled with the sub-zero drafts that rushed through gaps in the Flying Boxcar, made sleep impossible. Besides, passengers needed to maintain a vigilant eye on the altimeter hung overhead, lest oxygen masks be needed.

In this way, he endured not only the primitive discomforts but also the unpredictability of fellow travellers, including one who became so overwrought by the experience that he deployed his inflatable lifejacket while thrashing about in panic. The disoriented man then failed to fit his oxygen mask during a toilet visit and promptly passed out, leaving Oliphant and another passenger to dash to his rescue and drag the barely conscious man back to his mattress. At that point, the rip-cord of the patient's parachute snagged and a huge silk sheet filled the already cramped confines, billowed by the icy winds that whistled though the darkness.

* * *

When it came to the manifest details of personnel being shuttled back and forth across the Atlantic, Oliphant appreciated the non-negotiability of secrecy provisions. Yet in other respects, the adherence to absolute confidentiality that Groves both preached and practised offended Oliphant's belief in the unhindered exchange of scientific knowledge. And as the Manhattan Project grew in size and intricacy throughout 1944, this would increasingly land him in hot water, not unlike the heartburn his indiscretions generated during the radar program.

At one point, he came to the attention of General Groves via the transcript of an unauthorised telephone call made from an office at Berkeley, in which Oliphant discussed matters relating to the deployment of other British staff within the Manhattan Project. While that breach was scarcely the stuff of national security, it meant Oliphant would be closely monitored.

That became apparent to him soon afterwards, when he was returning from an evening stint with Lawrence at the giant calutron atop the hill overlooking the Berkeley campus. He approached his office to find a light blazing and a pair of plain-clothes FBI agents rifling through his filing cabinet, where he carefully locked away some of his more incendiary correspondence.

Oliphant's response to what he considered an inexcusable breach of his privacy was typically pragmatic. Come the next morning, he pulled apart the cabinet's lock in a reprise of his schoolboy tinkering, and reset the security combination without advising the intelligence agency he had done so.

In mid-1944, Groves further vented his anger about information that Oliphant had shared based on RAF surveillance photographs of a possible atomic production facility in Germany. By that stage of the war, information flowing to the Allies on the Nazis' uranium fission progress had virtually dried up, partly due to Groves's secrecy obsession. The general was not prepared to let intelligence officers on the ground in Europe know what an isotope separation plant or bomb research site might look like, for fear those men might be captured by the enemy and their knowledge extracted.

As a consequence, the information gleaned from studying aerial photographs was tightly held, until Oliphant was shown some of that pictorial evidence and ventured his opinion to colleagues on what it might show. Which brought an immediate scolding from Groves, and an equally combative response from the Australian.

'I would like to say that I am rather tired of these veiled accusations of carelessness or evasion of security rulings,' Oliphant wrote to Groves, outlining the reasons for his actions and the steps he had taken to ensure he had observed what he considered to be sufficient security protocols. He finished his one-page note to the feared general with a flourish: 'These considerations seem to matter far more to some people than does getting the job done!'[8]

* * *

'Getting the job done' was, however, proving to be a slow and difficult task. The problems that bedevilled the electromagnetic technique for separating the elusive uranium isotopes were testing Groves' notoriously febrile temper, and frustrating the scientists and technical staff working around the clock to meet the goal of a bomb.

Oliphant was one of those who would routinely finish his working day well after midnight. Despite the gruelling hours and the constant stream of technical problems that arose, he remained an upbeat presence, with his hearty laugh – almost Rutherfordian in its rumble – distinctively resonating up and down the hallways and workspaces of Berkeley and Oak Ridge. His warmth of character and preparedness to match words with action meant he was held in high regard, even if some of the ideas he put forward lacked practicality. At one stage, in mid-1944, he advocated that almost every suitable scientist who continued to work in the field of nuclear physics in Britain should be transplanted to the United States to supplement the Manhattan Project.

His indefatigable zeal, fuelled by an unhealthy reliance on black coffee and cigarettes, stemmed partly from his love of working on intricate machinery such as the calutron, and partly from the keen sense of camaraderie he felt with Lawrence and his team. But just as influential was his nagging fear that, even if the war was beginning to turn against Hitler, Germany's atomic research ambitions were ongoing.

It was that shared urgency that saw more British researchers enlisted for the Manhattan Project, including Philip Moon, who joined Oppenheimer and Peierls at Los Alamos. It meant that the Tube Alloys program in Britain essentially ground to a stop as laboratories like Birmingham and the Cavendish saw most of their nuclear physics staff relocate to the United States.

However, the increase in personnel was not mirrored by a similar upsurge in the production of the crucial uranium-235 at Oak Ridge, as further breakdowns and operational issues dogged the various isotope separation plants. Oliphant continued to call for an expansion of the electromagnetic facility in an effort to increase its output, but again he found no support among the upper echelons of the bomb program.

Instead, Groves was turning his attention towards an altogether different solution.

* * *

Another option as source material for a 'super-bomb' – an option the original MAUD Report had dismissed, but which had since been shown to carry significant potential – was the synthetic element 94. It had been discovered during work conducted with Ernest Lawrence's 60-inch cyclotron at Berkeley, and found to possess a basic chemistry that was very similar to uranium, from which it had been derived.

When the predominant uranium-238 isotope was bombarded by neutrons slowed down by a moderator, it was found to transmute into uranium-239. This isotope then rapidly transformed into a new unstable element of atomic number 93, which does not exist in nature. After a few days, it decayed into another synthetic product, the stable element 94.

Just as uranium – element 92 – was named in honour of Uranus, the seventh planet from the sun, after being discovered in 1789, these manufactured elements that followed on the periodic table took on the equivalent solar system sequence. Element 93 became neptunium, and 94 plutonium.

The latter was found to be even more fissionable than uranium-235, and released a greater store of neutrons when it split. It also boasted a critical mass smaller than the difficult to extract uranium-235, and therefore offered the decided advantage that less of it would be needed to fuel a nuclear weapon.

When clandestine word reached Oliphant – through the supposedly impermeable silo walls of the Manhattan Project – that plutonium was being viewed as the preferred core for the war-ending bomb, his disillusionment was palpable.

Three nuclear reactor 'piles' were now taking shape within a 2000-square-kilometre site on the Columbia River at Hanford. If the plutonium they produced became the heart of bomb plans, then the years of work Oliphant had dedicated to uranium separation would prove wasted. Given the problems that had plagued the electromagnetic plants under his direction, while Lawrence continued his work back at Berkeley, the lack of support for his push to have more resources allocated to them, and now the possibility they might even be mothballed in favour of reactors churning out plutonium, Oliphant's previously unsated appetite for the project began to wane.

Not that his scepticism was rooted purely in the threat of repudiation of his area of expertise. The imperative that had drawn him into military work was to devote whatever scientific insights he possessed to ending the hellish war. And it was doubts about the unproven weapons capability of plutonium that worried him most.

Those reservations were laid bare in a memo he sent to Lawrence from Washington in mid-February 1944, spilling details that he had absorbed from confidential conversations with Chadwick and Niels Bohr, now that the Dane had also joined the Manhattan Project's ordnance research team at 'Y' in Los Alamos.

> Both … [Bohr] and Chadwick tell me that at 'Y' they are very pessimistic about ever getting sufficient [uranium] 235 for a weapon and hence, are giving all their attention to 94 [plutonium]. It is this assumption that leads to all the ordnance difficulties … there are practically none with 235. Chadwick had endeavoured to persuade them that there is at least some chance of 235 being available, but he says they are only half convinced. You will appreciate that the information which I have given you about things, which I have heard here, is 'off-the-record' but I feel it essential that you should have it and that you should draw your own conclusions.[9]

As had been established from the time of Frisch and Peierls's memorandum three years earlier, once a critical mass of uranium-235 was achieved, the detonation of an atomic weapon was comparatively simple. The two sub-critical components were kept safely separate within the bomb to prevent premature detonation, then fired together using a gun-like mechanism at the desired moment, setting loose the chain reaction. And annihilation.

Although the critical mass needed for a plutonium weapon was much smaller – around one-third the amount of uranium-235 – triggering an explosion would be infinitely more complex. The highly fissionable plutonium core was not suited to 'gun' detonation due to the high risk of untimely explosion.

Following furious theoretical calculation and investigation at Los Alamos, it was agreed that implosion was the most feasible way to achieve ignition. The sub-critical masses of plutonium would be surrounded by high explosives which, when set off, would force the two sections of the core violently together to trigger the fission reaction. It was a method so speculative that it would require test firing before any weapon could be considered suitable for use in combat.

Oliphant's unease was therefore not primarily caused by questions over the bomb's key ingredient. Most of his anxiety stemmed from his pragmatist's view that, once completed, the device must work. The risk posed by a prematurely exploding nuclear bomb was unconscionable. And the repercussions of dropping a fully formed device on enemy territory and having it land harmlessly intact, as a gift-wrapped make-your-own-bomb kit, were potentially worse.

As it turned out, Oliphant's misgivings were shared by Oppenheimer, who was now having to deal with the realities posed by the highly enriched plutonium beginning to emerge from the Hanford reactors. As a synthetic element whose existence had been known for barely four years, its chemical, physical and metallurgical properties remained largely a mystery.

The complexities of designing a safe (to the deployers) means by which a plutonium bomb might be detonated even led Oppenheimer to consider resigning his commission as director of the project's main site. However, that outcome was averted through the application of the program's definitive feature and enduring strength: its sheer volume of willing personnel.

As the Los Alamos facility's technical history would note: 'The Laboratory had at this time [mid-1944] strong reserves of techniques, of trained manpower and of morale. It was decided to attack the problems of the implosion with every means available, "to throw the book at it".'[10]

To bring the vision of a plutonium bomb to fruition, the number of full-time staff in the Los Alamos laboratory, which had been around 1200 in mid-1943, would double, and then double again by the end of 1944.

As the Manhattan Project continued to grow exponentially into the largest science endeavour the world had known, the other unstinting constant was the obsessive secrecy in which Groves enshrouded it. When President Franklin D. Roosevelt died suddenly in April 1945, with the weapon nearing finalisation, his successor Harry S. Truman came to office with no inkling that an atomic bomb project had even been in operation for more than three years.

By that time, plutonium production had increased after nuclear fission had become self-sustaining at the largest of the three reactors at Hanford in September 1944. At Chalk Hill, under John Cockcroft's leadership, Canada's first nuclear reactor was also taking shape. And at Oak Ridge, the three different methods of uranium isotope separation that had been effectively competing against each other were now operating in harness to churn out low-grade uranium-235 that could then undergo further enrichment in Lawrence's calutrons. It was over these racetracks that Mark Oliphant continued to maintain a watchful eye, despite his broader unease.

And it was during those long, late-night 'owl watch' shifts at Oak Ridge that his restless mind began to turn to other things.

* * *

As he had shown through his earlier endeavours, Oliphant needed to be where the action was. The reason why he had decided, at age twenty-three, to pack his academic life and young wife off to England was to join Ernest Rutherford's atomic revolution. He had been central to the golden era of Cambridge physics that had followed, then had sought to engineer his own at Birmingham after Rutherford's death.

He had overseen the radar innovation that had helped Britain endure its 'darkest hour', and through sheer force of character had rallied American scientists in their commitment to the bomb. But now, after his belated introduction into the Manhattan Project and the dilution of his influence upon its secretive, labyrinthine structure, Oliphant felt himself to be on the periphery.

Ultimately, it was his allegiance to Lawrence and to seeing the project through that kept him in the United States throughout 1944. He remained crucial to trouble-shooting at Oak Ridge's electromagnetic plants, and he revelled in the incremental success of each team milestone. But by the beginning of 1945 – as a completed bomb came closer with every enriched atom painfully extracted from uranium in Tennessee, and every microgram of plutonium from the high desert of Washington State – he began to feel his continued usefulness to the project was limited.

Never one to linger when he was not at the heart of the action, Oliphant eventually decided his time with the Manhattan Project had come to an end. On 22 January 1945, he wrote a succinct, one-page memo to General Groves, advising his plans – already communicated to Akers and other British authorities – to return permanently to Birmingham in mid-March.

Chadwick, upon seeing its detail when it was shared by Oliphant as his direct subordinate, sent an immediate follow-up to Groves in an attempt to mitigate any ill will the premature departure might spark. Chadwick was fully aware of Groves' lingering suspicions that British scientists would use the information gleaned from their involvement in the Manhattan Project for the commercial benefit of Tube Alloys.

> Oliphant's position is not so uncompromising as it is put …
> Oliphant says he has urgent reasons not connected with T.A. [Tube Alloys] which demand his return to England. He believes also that he cannot contribute much more to the E.M. [electromagnetic] work as far as the war effort is concerned, and he is naturally anxious to be where he can be most useful.[11]

Chadwick's official correspondence with Groves was suitably conciliatory, but from the private conversations he had held with Oliphant – via letter and telephone, and when their paths occasionally crossed in Washington – his opinions differed. That's because, in his own mind, James Chadwick was beginning to feel similarly vexed. The health problems he had endured since his internment at Ruhleben ensured he appeared forever gaunt and stern, and his workload throughout the bomb project – compounded by his almost daily dealings with the combustible Groves – meant he appeared older than his fifty-three years as the war entered its final phase. As his good friend and regular confidant mulled over plans to return to England, Chadwick assured Oliphant he retained his full support.

Given the innate understanding and fondness they had developed for one another since their first meeting at the Cavendish almost two decades earlier, Oliphant knew that Chadwick shared some of his own doubts but, unlike the Australian, was too habitually diplomatic to air them publicly.

> An old-fashioned British patriot, Chadwick was wholly devoted to all that the United Kingdom represented. He admired and liked many Americans, but he did not trust their political, system nor its 'way of life'. He felt strongly that the nuclear weapons could not be left in American hands alone, and it was Britain's duty to possess the weapons as soon as possible, and thereby to have at least one sane and experienced nation with some say in its development and use.[12]

In Oliphant's mind, there was no equivocation. The bomb that had for so long been an experimental exercise was now glaringly imminent. And the project to which he had been personally attached since the Frisch–Peierls Memorandum landed in his hands was now grinding inexorably to its climax. And whether he remained in the United States or returned to his laboratory at Birmingham, that same sobering outcome would be delivered.

He was not interested in personal wealth or scientific status. Instead, he wanted to utilise his now vast knowledge of nuclear physics to benefit Great Britain.

* * *

As foreshadowed in his note to Groves, Oliphant returned home to his family and to his Birmingham office in late March 1945.

His first priority upon resuming the Poynting Chair of Physics was to reclaim the laboratory that he had not occupied for almost two years. The vision he had formed during the 'owl watch' stints at Oak Ridge was to lead the development of a large-scale electromagnetic plant in Britain.

But before he could devote himself to that ambition, he needed to restore some order within the physics department.

In his absence – and that of his former colleagues who had also joined the Manhattan Project – his laboratory had been occupied by a team from Oxford University. They formed part of what remained of Tube Alloys. Before he had departed the United States, Oliphant had demanded of Tube Alloys boss Wallace Akers that the interlopers be evicted from Birmingham as a priority. He had also advised his laboratory would no longer be available to Tube Alloys, given its threadbare form. Emboldened by Chadwick's reassurances, he foresaw that Britain's nuclear industry would thrive in the post-war world and therefore imagined Tube Alloys in its current moribund state would be subsumed into a much larger entity.

It was an exchange that prompted angst and ill feeling at senior levels of Tube Alloys and within the British bureaucracy before Oliphant got his way. As Vice-Chancellor Raymond Priestley was moved to note in his diary: 'Oliphant and the Admiralty Research Department have parted brass-rags and there is a great deal of feeling unfortunately. The more brilliant the scientist, the more gently he has to be handled.'[13]

* * *

Even though Oliphant's attention was turning to Britain's role in the peacetime nuclear landscape, he remained an enthusiastic advocate for the Manhattan Project and the impending realisation of its mission. As he saw it, the completion of an atomic bomb would bring the exclamation mark needed to end the ongoing war. With the same display of harrowing force, it would also announce a new era of peace, in which the full potential of nuclear energy could be harnessed for the betterment of civilisation.

Shortly before leaving for Birmingham, he had written an optimistic parting letter to Lawrence:

> There is no doubt we have been associated with the birth of a new industry. This is not a passing phase in the intensive development of

a new military weapon but a permanent contribution to science and technology of the future.

It will not be long before we see demonstrated the first fruits of our labours. Though war has brought the opportunity to do these things, and although the immediate result will be incalculable destruction, we know that in the ultimate analysis this aspect will be overshadowed by the benefits wrought for mankind.[14]

The letter gave no indication of whether he anticipated that the bomb's 'incalculable destruction' would be wrought, as he hoped, upon the planet itself, or unleashed directly on its people.

Having settled back in Britain, his final correspondence with Groves on 3 July 1945 also carried no trace of misgiving as to how the United States military might choose to use their new super-weapon.

'May I congratulate you also on having brought to fruit so great and so novel a venture,' Oliphant wrote. 'If the imminent next step proves as successful as I believe it must, we will see a complete vindication of the faith of those of us who have fostered this revolutionary undertaking, and incidentally a great demonstration of the practical value of academic nuclear physics.'[15]

Thirteen days later, that 'imminent next step' was irreversibly taken.

22

'DEATH, THE SHATTERER OF WORLDS'

United States and Japan, 1945

At 5.30am on 16 July 1945, upon the barren desert stillness of Alamogordo Army Air Base, the world's first plutonium bomb lit the pre-dawn New Mexico horizon with a blazing, faux sunrise that announced the nuclear age.

Delayed ninety minutes due to poor weather, the test designed to prove the device's implosion detonation mechanism was codenamed Trinity. Its success led Oppenheimer to famously recite a line from the Hindu scripture the *Bhagavad Gita*: 'I am become Death, the shatterer of worlds.'

For members of the Los Alamos laboratory's co-ordinating committee – who had been bussed to a viewing site thirty kilometres from ground zero and provided with a sheet of welder's glass as they lay prone on the cold earth – it brought a perverse wonder. For an agitated few, it also ushered in a wave of comfort amid their genuine dread that 'big science' might have grotesquely over-reached.

As early as 1942, as Oppenheimer's small group of theorists met at the ridge-top compound to brainstorm the latest apocalyptic insights, Hungarian physicist Edward Teller had brought the gathering to silence when he mused that a nuclear fission bomb might feasibly set the entire planet ablaze. That such an unprecedented, unknown surge of energy could potentially ignite the flammable nitrogen in the earth's atmosphere, and even set fire to hydrogen in the oceans – ending all planetary life in a blinding inferno.

'The earth would blaze for less than a second in the heavens and then forever continue its rounds as a barren rock,' former United States military advisor Daniel Ellsberg reckoned of that plausible threat.[1] Following Teller's macabre scenario, another of the group's émigré scientists, German Hans Bethe, had gone away to calculate the probability and found it 'extremely unlikely to say the least'. It was not, astonishingly, *absolutely* zero, despite Oppenheimer's post-war pronouncement that 'the impossibility of igniting the atmosphere was ... assured by science and common sense'.[2]

Not everyone at Los Alamos was so easily placated, however. On the eve of the Trinity Test one of the theorists, Robert Serber, met Teller pacing the windy, dust-strewn streets of Los Alamos. The Hungarian asked Serber how he planned to counter the threat of rattlesnakes, which Oppenheimer had warned might be lurking in the dark while his team lay on the desert floor, awaiting whatever the bomb would bring.

'Well, I'll take a bottle of whiskey,' Serber answered facetiously. As he wrote years later: 'Then he [Teller] brought up the notion that the atmosphere might be set on fire by the bomb and he said, "What do you think of that?" And I said, "I'll take another bottle of whiskey."'[3]

General Groves's thirteen-page report on the Trinity Test for United States Secretary of War, Henry Stimson, evaluated the exercise as 'successful beyond the most optimistic expectations of anyone'.[4] 'The Gadget', as it had become known, was released not from an airborne bomber, but from a thirty-metre-high steel tower that vaporised when the plutonium core, conservatively estimated to equal the force of 20,000 tons of TNT, fissioned in a millionth of a second. It blew a crater more than 350 metres in diameter into the earth's surface. When he saw the damage it wrought upon another steel tower one kilometre from the drop zone, Groves knew that not even his prized Pentagon building would offer safe shelter from such a weapon.

Oliphant was taking lunch at Birmingham University when the first nuclear bomb was detonated in the chill of the New Mexico dawn.

Groves's censorship powers had ensured that the inevitable news inquiries would be met with an official four-paragraph statement, disingenuously attributing the dazzling effect of the monstrous fireball – visible up to 300 kilometres away – to 'a remotely located ammunition magazine containing a considerable amount of high explosives and pyrotechnics [that] exploded'.[5]

However, Oliphant's friend Ernest Lawrence had stood barely forty kilometres from the point where history erupted, and later that day wrote an account of what he saw.

> Through my dark sunglasses there was a gigantic ball of fire rising rapidly from the earth – at first, as brilliant as the sun, growing less brilliant as it grew boiling and swirling into the heavens. Ten or fifteen thousand feet above the ground it was orange in colour and I judge a mile in diameter. At higher levels it became purple and this purple afterglow persisted for what seemed a long time (possibly it was only for a minute or two) at an elevation of 20–25,000 feet.[6]

Oliphant, like Lawrence, would have understood the violet glow preceding the violent roar to be the radioactive discharge within the surrounding atmospheric gases. It was an effect they had often seen on a vastly smaller scale within the high-energy cyclotron. When Oliphant learned the details of the Trinity Test from sources within the Manhattan Project, he also knew what that success meant for the bomb project he had so devotedly helped nurture to nativity.

'Once the test in the desert took place in New Mexico and it was clear that all our fears were justified, it then was no longer the concern of the scientists,' Oliphant would reflect almost fifty years after the Trinity Test. 'It went straight into the hands of the military and the politicians. Scientists had nothing more to do with it.'[7]

* * *

In the Pacific, the atrocities of conventional war mounted even as Britain rejoiced at Germany's defeat in May 1945. In March, a single United

States firebombing raid on a section of Tokyo, densely packed with timber-and-paper houses, had killed more than 100,000 men, women and children, and injured a million more in the space of six hours. Yet Japan's leaders refused to yield.

Winston Churchill was at Potsdam in newly occupied Germany, meeting with United States president Truman and Soviet premier Josef Stalin, when he learned of the Trinity Test. Britain's leader immediately announced that anything other than unconditional surrender from Japan would compel Allied forces to 'conquer the country yard by yard [and] might well require the loss of a million American lives and half that number of British – or more'.[8]

Then, just days after the release of the Potsdam Declaration, which included terms to which Japan must agree or face 'prompt and utter destruction', the heavy cruiser USS *Indianapolis* was attacked and sunk by a Japanese submarine, while travelling between Guam and the Philippines.

The *Indianapolis* had only recently arrived in the Pacific from San Francisco, and four days earlier (on 26 July) had delivered the atomic bomb's gun assembly mechanism to its destination. Of the more than 500 crew who died, those who did not burn in the flaming fuel slicks that spread across the Philippine Sea, or were not entombed in the ship's steel carcass as it plunged bow-first to the ocean bed, most were taken by sharks during the four days they waited for rescuers to arrive. That rescue mission was mounted on the morning of 2 August 1945, which would have been the day after the first atomic bomb was dropped had a typhoon approaching Japan not forced that plan's delay.[9]

A further two days later, on the sugarcane-covered Northern Marianas island of Tinian – which had been seized from the Japanese a year earlier and was barely wide enough to support a full-length military runway – the events Mark Oliphant so feared were unfolding, under secrecy's heavy cloak.

The sixty-four kilograms of concentrated uranium-235 collected from Lawrence and Oliphant's calutrons at a rate of around 250 grams per day over the previous months were now packed inside a four-ton bomb known as 'Little Boy'. It was different in both shape and science from the plutonium bomb, a prototype of which had been successfully tested in the New Mexico desert the previous month.

In line with the codenames attached to all other elements of the Manhattan Project, the two types of bomb were disguised in Air Force communications to give the impression that their cargo was VIP passengers rather than precious ordnance items. Thus 'Thin Man' – which evolved into 'Little Boy' – suggested Roosevelt, but was instead a reference to the uranium bomb. Its larger plutonium cousin became 'Fat Man', representing Churchill.

The culmination of five years' pioneering science, complemented by human endeavour on a scale considered civilisation's largest single-focus enterprise since the construction of the pyramids, Little Boy was essentially a steel-encased cannon. Inside it, explosives would fire a bullet of uranium-235 down a short barrel, at the end of which it would slam into three rings of the enriched isotope.

The bounty of a process that had – as Bohr predicted – required the resources of an entire nation resembled 'an elongated trashcan with fins',[10] according to a crew member, as it was loaded gently into a B-29 Superfortress bomber, freshly emblazoned with the given names of pilot Paul Tibbets's mother, Enola Gay.

* * *

At 8.15am on 6 August, on a flawless summer morning in Hiroshima, hell appeared out of the cobalt-blue sky. The bomb, inscribed with messages from United States ground crew that included 'Greetings to the Emperor from the men of the *Indianapolis*', detonated – by means of an airborne radar unit adapted to engage the gun mechanism at a predetermined altitude – 500 metres above the city, which housed a sizeable army depot.

Souls were seared into the streetscape. Buildings were reduced to memories beyond dust. Passing birds were vaporised in the blazing sky.

As had been foreseen by the men of the MAUD Committee, the gun method of initiating the catastrophic chain reaction worked without a hitch. What was not known until the first – to date, the only – atomic bomb was set loose was that only around two per cent of its uranium core underwent fission before the evil consumed itself and blew apart the remainder of the enriched material in the burning air. That was sufficient to replicate the effect of almost 12,500 tons of TNT.

Of the 76,000 structures in Hiroshima, around 70,000 were damaged. It was only pre-emptive fear of possible Allied bombing, which had led to large-scale evacuation from the city in previous weeks, that had reduced its population from 400,000 to less than 300,000 that Monday. Of that number, around 140,000 died as a direct result of the blast. Those who weren't instantly incinerated subsequently succumbed to horrific burns.

Three days later, 'Fat Man', an even more powerful plutonium device, exploded above Nagasaki, the city where the torpedoes dropped on Pearl Harbor had been manufactured.

The Japanese Empire's unconditional surrender came less than a week afterwards, on 15 August 1945.

23

'WE HAVE KILLED A BEAUTIFUL SUBJECT'

North Wales, Birmingham and New York, 1945

The age of atomic annihilation might have arrived out of flawless blue sky above Hiroshima on a busy Monday morning, but in the United Kingdom it was August bank holiday weekend. And the Oliphant family were on seaside vacation in North Wales.

Like many who rambled along Rhyl's carnival shorefront and braved the torpid water, which kept a sharp chill despite heavy humidity, seven-year-old Vivian Oliphant was basking in freedom. The first summer season after six years of wartime privation, with its airborne night terrors and restrictions on food and comfort and normality, meant that, for most holidaymakers, the long weekend stretched even longer – into Tuesday and beyond.

In that euphoric season following Germany's defeat, many of Rhyl's waterfront hotels and private boarding houses had been requisitioned for use by Britain's army and civil service, and subsidised rentals offered as reward to personnel who had helped repel the Nazi threat. It lent the careworn coastal community a gala veneer as victory celebrations entered their third month.

British rail officials reported the train network's heaviest bank holiday patronage in twenty-five years. One group of unfortunate travellers were prevented from disembarking at their intended destination due to the crush in their carriage, and were forced to alight at the next stop, almost 480 kilometres up the line in Scotland.[1]

Signs that hardships were lifting like morning fog from the Irish Sea were obvious along Rhyl's beachfront. During daytime, families queued for pantomime performances at the open-air Coliseum Theatre. Ethel 'Sunny' Lowry, the local heroine who twelve years earlier had secured stardom as the first British woman to swim the English Channel, presented her wildly popular aquatic shows at the outdoor baths. Even the town's landmark pavilion – with its ornate domes still swathed in camouflage cloth as a legacy of the days when German bombers stalked the breadth of the British Isles – hosted a slate of new stage plays.

On that cloudy and muggy Tuesday morning, however, freedom for Vivian Oliphant – whose only conscious memories were of life during wartime – took the form of stolen escape from the hawk-like vigilance of her parents. For reasons that had escaped Vivian's notice, her mum and dad had opted to retreat to a nearby café while she skipped her way to the waterfront.

If Vivian sensed something was awry in her family, it was not sufficient to dissuade her from tackling her first-ever solo ocean swim. A slight, dark-haired girl who had not been taught the essentials of water safety, Vivian seized her rare moment of independence and waded uncertainly into the shallows, which lapped lazily at the coarse brown sand.

But her optimism slipped away as suddenly as the ocean floor disappeared underfoot. Her arms flailed in panic above her head, plunging her face beneath the now-frothing surf. Brine flooded her nose and throat, further gagging her fitful peals of distress.

Her guttural cries for help were lost among the languid squawks of Atlantic gulls gliding above, while around her groups of revelling Britons squealed and laughed, splashed and frolicked. Blissfully oblivious.

For the rest of her days, Vivian would have no clue as to the identity of the teenage boy who saw her struggle and saved her from drowning. She knew only that it was not her brother Michael, who had been unaware of the emergency. Nor could she recall how she was helped ashore, where beachgoers crowded around her in a flap.

'WE HAVE KILLED A BEAUTIFUL SUBJECT'

Disoriented by her trauma, Vivian could only assume that someone from that swarm had found Mark and Rosa Oliphant at one of the nearby tea houses. But her understanding of why her dad and mum had been so uncharacteristically negligent would soon become clear, and that memory would remain with her, pin-sharp, forever.

That same morning – Tuesday, 7 August 1945 – Britain had awoken to details of the atomic bomb that had unleashed hellfire on Hiroshima. And of Mark Oliphant's central role in delivering that atrocity.

* * *

It was early Sunday evening in Rhyl when *Enola Gay* throttled down the pitch-dark runway and rose, groaning, into infamy. She had begun her twelve-hour, 5000-kilometre round trip from Tinian to Hiroshima at the appropriately ungodly local hour of 2.45am. On the following evening, the Oliphants took their holiday-Monday dinner at the prescribed time, after which Michael and Vivian, worn out by a day of paddling and playing, fell quickly asleep.

Having been made aware of the United States Army's successful nuclear bomb test in the New Mexico desert three weeks earlier, Mark Oliphant suspected it would not be long before another device was detonated, in a bid to end the ongoing war with Japan in the Pacific. Not that he spoke those misgivings aloud, to either his wife or his children.

He was, however, one of eight British scientists involved with the atomic bomb program to have received a telegram from Wallace Akers on Saturday, 4 August 1945.

> It is possible that an official announcement on your project may be issued any time from now on. You are likely to be approached by the press as the names of the members of the technical committee will be given. You should state that you regret that you are not permitted to give any information whatever and that you are advised that a fuller official statement will be released very shortly to the press.
>
>   You must not allow yourself to be drawn into any discussion on the underlying scientific phenomena as the fuller statement will

include a most carefully composed historical survey and it is very important that this should be dealt with as a whole. You may of course give biographical details about yourself if asked. I suggest that you should get in touch with me by telephone on Tuesday to discuss further arrangements. Akers.[2]

Unfortunately, by the time the telegram boy cycled to Peto that Saturday afternoon with the cryptic communiqué in hand, the Oliphant family were already enjoying the seaside delights of Rhyl. So the cable remained, uncollected, at Barnt Green's post office.

Instead, the news was delivered by the radio set around which Mark and Rosa sat on the following Monday night. At that point, the weapon's true impact still remained mostly unknown, due to the vast cloud of smoke and dust that had followed the fireball when it exploded in the dead of the previous Welsh night. The gruesome detail of the bomb's effect, therefore, would emerge only in the ensuing weeks, months, years.

However, the enormity of that moment and its immediate relevance to the Oliphant family were conveyed that evening by a BBC newsreader, who reported with sombrely patriotic pride the 'tremendous achievement of Allied scientists [in] the production of the atomic bomb'.[3] He read from a 1500-word pre-drafted statement, released by British Prime Minister Clement Attlee but attributed to recently toppled wartime leader Winston Churchill.

That document gave a bureaucratic overview of the bomb's evolution, from abstract theory to manifest obliteration. It catalogued the crucial discovery of uranium fission in the years before war, the vital role of British science in applying that knowledge while at the same time fighting for its very existence against Nazi Germany, and the unprecedented melding of human intellect and manufacturing muscle that finally brought the weapon to fruition in the United States.

The moment that caused greatest discomfort for the Australian couple, as they sat silently in their beachside apartment, was the roll call of British experts so prominent in this awesomely savage moment of history. With 'Professor Oliphant' among those singled out for laudable mention.

During that weekend, Rosa had noted her husband's edginess as hourly new bulletins neared, and the closer than usual attention with which he scoured the morning newspapers, but she had remained oblivious to the extent of the tumult he harboured. Despite his litany of indiscretions, Mark Oliphant had successfully withheld from his wife all information about his role in the development of this definitive killing device.

'He doesn't talk about his work,' Rosa would recall when interviewed months after the 9pm bulletin confirmed Mark's involvement in the bomb's production. 'Naturally, it's too important. I knew, of course, that he had been working on atoms, but I didn't know how successful it was until the news was officially released. It was a terrific surprise.'[4]

There was also alarm as she watched the colour drain and the usually sunny expression fade from Mark's cherubic face as the wireless set spoke of the force unleashed on an unsuspecting Japanese city. They were characteristics that remained absent from his visage when the couple rose next morning, after a fitful night's sleep.

The same text as had been read on radio was then laid out in the newspapers. Mark Oliphant's preferred broadsheet, *The Times*, pointedly printed 'The First Atomic Bomb' alongside its emblematic front-page masthead. Beneath the ensuing report stood four decks of headline in large capital letters:

FIRST ATOMIC BOMB HITS JAPAN
EXPLOSION EQUAL TO 20,000 TONS OF T.N.T.
ANGLO–US WAR SECRET OF FOUR YEARS' RESEARCH
'RAIN OF RUIN' FROM THE AIR[5]

Among the three columns of densely packed newsprint, along with the names of Oliphant and other scientists, was the deposed prime minister's declaration that 'by God's mercy British and American science outpaced all German efforts [to construct an atomic bomb]. These were on a considerable, scale, but far behind.'[6]

As Mark Oliphant would spend the subsequent fifty-five years of his life explaining, the effort he had invested in the atomic bomb was founded on a single burning belief: that the genocidal doctrine pursued by Adolf Hitler could not be allowed to succeed. The consequences if Germany secured the bomb outweighed any concern as to how it might be used by the Allies. Fused with the terror and devastation he had seen wrought upon Britain's cities through nightly Nazi bombing raids, Oliphant's antipathy towards Germany had fermented into unconcealed hatred.

What Oliphant, and others, would not know until long after the war concluded was that Nazi Germany was even further from mastering nuclear technology than Churchill's statement had suggested. That reality had been confirmed to intelligence agencies through the capture and internment of Germany's foremost physicists, who were secretly held at a Georgian manor not far from Cambridge University following the Allied victory in Europe.

Every room at Farm Hall, the walled residence in semi-rural Godmanchester where the renowned scientists were kept under house arrest, was fitted with listening devices. Through the elaborate surveillance program, each fleeting utterance and shared conversation was recorded, laboriously translated into English and cabled on a daily, highly confidential basis to General Leslie Groves in Washington.

Among those who were held captive for six months were two Nobel Prize winners, and a third – Otto Hahn, discoverer of nuclear fission – would learn of his Nobel honour during internment. Another of the ten inmates was Paul Harteck, the man who had worked shoulder to shoulder with Oliphant and Rutherford in the discovery of tritium at the Cavendish Laboratory before returning to Germany.

But the true moribund state of Germany's bomb program was not fully understood until the unguarded astonishment that was betrayed by the men when they learned of Hiroshima's devastation. The news was relayed to them via the same 9pm BBC radio bulletin that so troubled Mark Oliphant.

As that Monday evening progressed, and the scientists nearest to Germany's bomb program numbed their shock with house wine, their raw reactions confirmed to Groves that the Nazis had been nowhere near the successful completion of an atomic weapon. They could not hide their disbelief that the Allies had finished a functioning bomb in such a timeframe, and some dismissed the news report as a propaganda hoax.

It would be decades before edited sections of the Farm Hall transcripts were publicly released, but by then details of the Germans' attempts to build a bomb had leached out. The mutual mistrust between Hitler's government and many of the senior scientists involved in the project, as well as the defection of so many crucial scientific personnel to Germany's wartime rivals, meant the Third Reich had not even managed to develop a functioning nuclear reactor by the time it was defeated. Given that the Japanese never seriously embarked on a nuclear weapons program, it became clear to Oliphant that his relentless drive to secure the bomb before Britain's enemies did so had been forged upon a deeply flawed premise.

* * *

By the time the second bomb was dropped on Nagasaki, three days after the destruction of Hiroshima, the Oliphants' much-awaited family vacation had been cut short. Realising the device he had doggedly believed would curtail the atrocities of war had instead been used to escalate them, Oliphant had rushed back to Birmingham.

On arrival at his office, he found the public release of Churchill's statement had already endowed him with a measure of celebrity. Within a day, he was approached by Hugh Warren, managing director of heavy manufacturing enterprise British Thomson-Houston, to deliver a presentation to the firm's engineers: 'now that the lid is off – exposing the top layers [of the bomb project]'.[7]

As Sir Hugh noted in his invitation, there was a general feeling that Oliphant had received 'extremely scanty mention' in official summaries of the nation's atomic contributions. That was probably because he had not been among the core of scientists working under Oppenheimer, the bomb's acknowledged 'father', at Los Alamos. Plus, he had disentangled himself entirely from the Manhattan Project months before its deadly secret became known. But that did not assuage his own sense of culpability as reports of the horror in Japan became increasingly graphic.

His first public response to the barbarity he had helped inflict came in the weeks after the dual strike on Japan. In the midst of the V-J Day celebrations, after Japan had formally signed surrender documents in early September, Oliphant took up a guest-speaking invitation with Birmingham's Rotary Club. There he claimed that, having used the bombs in the slaughter of innocents, the war's victorious parties 'cannot complain if one is eventually dropped on us'.[8] Newspapers in Britain and Australia carried details of that speech, in which his initial distress had clearly distilled into simmering anger.

At the heart of his fury lay his belief that Japan's submission might well have been achieved by staging an increasingly threatening series of deterrent strikes. It could have begun with the demonstration of the bomb's frightening potential at a remote location, with the result that human life and property would have been preserved. Then, in addition to an intensive propaganda campaign waged by radio and leaflet drops, a subsequent bomb could have been detonated off the Japanese coast to display its ferocity. If that were not effective, the next step could have been to destroy one of the Tokyo Bay islands utilised as a naval base, and only as a last resort should the device have been used on a city.

'I've always thought it was a tragic mistake to actually drop bombs on Hiroshima and Nagasaki. They should have been used to blow the top off Mount Fujiyama,' Oliphant would later reflect.[9]

However, this overlooked the reality that the bombs used on Hiroshima and Nagasaki were the only two in the United States arsenal at that time. Had the warning shots been ignored, the war would have continued for months, possibly longer, with countless more lives lost while uranium was painstakingly collected or plutonium extracted for further devices. In fact, the scientific panel formed to advise President Truman on nuclear matters – which included Oliphant's friends Lawrence and Oppenheimer – had considered the deterrent option a month before the Japan strikes, and decided the psychological impact would not be the same. They reasoned there was no alternative to direct use.[10]

Oliphant, however, continued to assert that, had control of the bomb remained with the scientific community that had brought it to life, it would never have been deployed against civilians. And that any odium directed at the scientists among the upper echelons of the Manhattan Project should be aimed, instead, at the politicians who had ultimately hijacked it.

While he would stoutly defend his involvement in the Manhattan Project, and proclaim he was 'proud to have been associated with the wartime atomic energy project, for the alternatives that Germany or Japan had first possessed the nuclear weapons are too dreadful to contemplate',[11] Oliphant's retrospective disdain for the decision to bomb Hiroshima and Nagasaki never diminished. 'I felt so utterly disgusted and upset by this use of the nuclear weapons against a civilian population,' he would later reflect.[12] 'During the war I worked practically the whole time on defence research. I worked then on nuclear weapons so I, too, am a war criminal.'[13]

'I think nowadays I would not work on a project like that. The trouble is when your country is at war, well, you give yourself completely to its defence.'[14]

In particular, his soul was scarred by John Hersey's heart-rending first-person accounts from Hiroshima, published in the *New Yorker* a year after the bombing. Much of Oliphant's personal correspondence from before 1957 has been lost or misplaced in vast university archives, but that edition of the magazine stayed with him, as a form of moral touchstone. His wordless markings on the edges of those yellowing pages denoted the most confronting revelations, as a ready reference should his conscience ever waver.

Among the delicately highlighted passages of Hersey's reportage that had awakened the world to the horror inflicted upon Japan was an excerpt from an eyewitness report. It had been submitted to the Vatican by Father John Siemes of Tokyo's Catholic university.

Siemes noted: 'The crux of the matter is whether total war in its present form is justifiable, even when it serves a just purpose. Does it not have material and spiritual evil as its consequences which far exceed whatever good might result?'[15]

It was a question that would resonate with Oliphant throughout his life.

There was also a further macabre irony, in that the pinnacle of Oliphant's work in experimental physics – the 1934 discovery that the collision between two light nuclei could yield a heavier atom – would prove integral to the development of humanity's most obscene weapon, the thermonuclear hydrogen bomb, which superseded its uranium and plutonium predecessors. This was a fusion device ignited by a smaller fission bomb that, when successfully tested in the 1950s, was shown to be hundreds of times more potent than the plutonium weapon that destroyed Nagasaki. Its development also ensured that the convoluted process by which material was gathered for the (far less powerful) atomic bomb was rendered immediately obsolete.

Oliphant's conflict would be compounded when the third iteration of hydrogen that he had identified and named at the Cavendish – tritium – was identified as a booster agent that was also crucial to detonating these thermonuclear bombs.

His lifelong anger towards nuclear weapons would be fuelled, in no small part, by guilt.

* * *

Oliphant would continue to preach against the omnipotent threat that came to be posed by these weapons in the war's wake. He embraced this task with customary wholeheartedness in an article he penned for London's *Illustrated News* barely a month after the conflict's formal end.

'We are faced with the alternative – suicide or cooperation,' Oliphant noted of the race for nuclear superiority that loomed. 'Without a solution of the problem of war, enjoyment of the fruits of atomic energy will be, for the world, but the feast of the condemned before the execution.'[16]

His second, inter-related pursuit was to advance the case for nuclear energy as the answer to Britain's – indeed, the world's – escalating post-war energy needs. Journalists warmed to his charismatic and enthusiastic advocacy of the nuclear industry. As Britain enforced harsh electricity rationing due to coal shortages, and fears grew that its unfolding fuel crisis might reduce its standing as a world power, Oliphant had no doubts where the answer lay: in the vast source of energy that could be liberated from uranium atoms, and in employing the knowledge British scientists had gleaned during the war.

What none of those scientists had been able to foresee was the breakdown in the nuclear relationship between the United States and Britain come the postwar peace. The supposedly binding agreement to continue sharing the knowledge acquired through both nations' (and Canada's) involvement in the Manhattan Project was revealed to have died along with its key signatory, Franklin Roosevelt.

Its detail, and that of the subsequent 1944 Hyde Park Aide-Mémoire, which reiterated that full collaboration should continue after the war, was unknown to Truman. Not only that, but the United States also revealed in 1945 that the physical document could not be found anywhere. It subsequently classified all nuclear data as 'restricted' and brought an end to bilateral co-operation. In April 1946, Truman advised Britain that despite the earlier agreement, he did not believe the United States was under any obligation to assist its ally in the design, construction or operation of an atomic energy plant.

Oliphant was especially incensed to learn that Winston Churchill (in the autumn of his first prime ministership in mid-1945) had told Britain's House of Commons that, in keeping with the Quebec Agreement, the United States was not looking to exercise a virtual monopoly over future uses of nuclear technology. As Oliphant understood it, not only had Truman decreed that the United States would withhold collaboration on nuclear matters, but that volte-face also applied to peaceful as well as military purposes. He was therefore stunned by Churchill's blatant untruth.

'This was the first time I had ever heard a Prime Minister tell a deliberate lie and it so shocked me that I could never regain my wartime regard for him [Churchill],' he wrote decades later.[17]

Not prepared to simply sit back as political debate on nuclear issues unfolded, Oliphant found an accomplice through whom he could exert direct influence. Captain Raymond Blackburn had served in the Royal Artillery Regiment, and in the 1945 general election that swept Churchill from office, he was installed in parliament as an outspoken Labour MP. Blackburn soon gained notoriety as a fierce critic of new prime minister

Clement Attlee's approach to nuclear energy policy, under which a revised accord was signed that significantly weakened the terms of co-operation between the countries, and led calls for international policing of nuclear weapons.

In November 1945, Blackburn infuriated Whitehall when he regaled the House of Commons with details of the confidential Quebec Agreement. In the ensuing speculation as to the source of Blackburn's information, it became known that Oliphant was the maverick MP's unofficial advisor. Blackburn's willingness to take up the cudgels on Oliphant's behalf fitted snugly with the Australian's view that Britain had been betrayed by restrictions on its post-war involvement in the development of nuclear weapons as agreed to by Churchill, and enshrined within the Quebec Agreement. Attlee had subsequently noted that, while the ratification of the original Quebec Agreement under Churchill in 1943 was an important achievement, the document itself did not constitute a formal treaty and he was (as prime minister) compelled to renegotiate its terms.

'What we are doing now,' Oliphant wrote of the decision to effectively sign over Britain's atomic sovereignty to the United States, 'gives only the impression that we are trying to muscle in on a racket we have been too dumb to develop ourselves'.[18]

* * *

An extraordinary missed opportunity the following year galvanised Oliphant's earlier mistrust for politicians into abject disillusionment. He was persuaded to act as technical advisor to Australia's delegation at the early sessions of the newly established United Nations General Assembly in New York. That group would be led by former federal attorney-general and external affairs minister Herbert 'Doc' Evatt. He had been installed as the assembly's chairman at the inaugural session, given Australia's alphabetic primacy among member nations.

The convoluted six-week proceedings – held in the days before simultaneous translation, which meant every speech was repeated over and over in each required language – made for torturous hours of inaction.

When proposals were made relating to the use of nuclear weapons, which at that stage resided solely in the grip of the United States, Oliphant was left further disenchanted – most notably with Evatt.

Soviet representative Andrei Gromyko responded to America's support for the status quo by arguing that existing stockpiles of nuclear weapons should be dismantled. This suggestion brought Robert Oppenheimer, engaged in a similar capacity to Oliphant alongside United States delegate Bernard Baruch, rushing across the chamber to speak with his fellow physicist.

'For heaven's sake get your boss [Evatt] to say something in favour of the Russian proposals because that is wonderful,' Oppenheimer enthused to Oliphant. 'I think that we should consider them very seriously, and I'll tell my boss that is what I feel.'

Oppenheimer went on: '[As a] matter of interest, I'll give you a bit of classified information. At the present time there are only three nuclear weapons in existence. It would be half an hour's work to take them apart ... If the whole proposal failed, it'd take us another half hour to put them together again. So we've got nothing to lose by considering very seriously the Russian proposal.'

Oliphant then anxiously waited for a break in proceedings to brief Evatt, who was seated with gavel in hand at the bench overlooking the array of international representatives.

'I've been talking to Robert Oppenheimer,' Oliphant urgently told Evatt when the chance arose. 'He and I both believe that the Russian proposal[s] should be considered very seriously. Will you please make a statement to that effect, that we should discuss them in detail?'

Evatt turned in clear irritation and shot back gruffly: 'No, no, no. Nothing of the sort, we might want to use them against [the Soviet Union].'[19]

The first opportunity for de-escalation of a nuclear arms race was thus dead on delivery, three years before the Russians developed a bomb. It was a stark and ultimately calamitous illustration of Oliphant's earlier misgivings about the scientists who conceived and delivered the bomb being marginalised by the military and politicians from the very moment of its birth.

* * *

On occasion over the decades that followed, some of Oliphant's rebuttals when queried about his active role in the Manhattan Project would appear somewhat disingenuous. His claim that 'many of us nuclear physicists associated with the development of the nuclear weapons in the United States and in Britain were unhappy at the end result, we always hoped that the thing wouldn't work and that we would be absolved'[20] seems sharply at odds with his furious advocacy of the bomb's development – as do his enthusiastic, congratulatory notes to Lawrence and Groves as the weapon neared completion. Given the insistence and the energy he had shown in getting the bomb project to that point, it is difficult to entirely accept his revisionist regret.

His post-Hiroshima viewpoint certainly aligned with the small but vocal group of Manhattan Project scientists – Niels Bohr most notable among them – who made it known as the bomb program neared completion that deploying it against enemy civilians was an act no civilised nation should countenance.

By that stage of the war, Nazi Germany had been defeated. Before that, the all-consuming race to unlock the secrets of the atomic bomb seemed to have blinded these more altruistic members of the scientific community to the realities of global warfare. Or perhaps they hoped that the war would be completed by conventional means before any protagonist had perfected the weapon, at which time it might become an all-powerful deterrent against any future multilateral conflicts, as it has – in the aftermath of the Hiroshima and Nagasaki atrocities – proven to be.

The other possibility is that Oliphant understood all along what the end result would be once he and his colleagues had solved the enormous scientific conundrum on which they had been set to work. And that it was only when he was faced with the unspeakable consequences of his endeavours that his conscience dictated he spend the rest of his life preaching regret.

* * *

Where he remained utterly consistent, however, was on the other question regularly posed about his unrelenting campaign to ensure that the complex theory of atomic weaponry became cruel reality. That being: what would Ernest Rutherford have thought had he lived into his seventies, and borne witness to the horrific events of the Second World War?

Oliphant would be forever haunted by what he believed to be the unspoken judgment of his scientific inspiration, his fellow Antipodean, his closest friend.

At the heart of those feelings was Oliphant's awareness, drawn from their years of close professional and personal connection, of Rutherford's disdain for warfare – as well as his deep fear that the power of the atom he famously split at the close of the First World War might be hijacked for military and political purposes.

'We, who worked with Rutherford and enjoyed his trust and friendship, often wonder what "the Prof" would think of this terrible prostitution of knowledge for which he, above all men, was responsible,' Oliphant would muse decades after Hiroshima.[21]

A hallmark of Rutherford's remarkable research career was his capacity to 'see' experimental results before they were achieved. His instinctive feel for atomic structure seemingly allowed him to know what he would find from the moment he began searching.

In delivering a speech at Manchester's New Islington Town Hall at the bloody height of the First World War, Rutherford visualised the potential of nuclear energy, but feared for its use in such a climate of conflict. Later he also expressed grave worry over the increasing threat posed by military aircraft, and their potential to terrorise defenceless civilian populations.

Oliphant would reflect on those thoughts in a dissertation on his mentor's enduring significance, delivered a century after Rutherford's birth:

> [Rutherford] said that scientists wanted to ascertain how they could release at will the intrinsic energy contained in radium and utilise it for our own purposes.
>
> It had to be borne in mind that in releasing such energy at such a rate as we may desire, it would be possible from one pound of the

material to obtain as much energy practically as from one hundred million pounds of coal. Fortunately, at the present time, we had not found a method of so dealing with these forces, and personally he was very hopeful that we should not discover it until Man was living at peace with his neighbours.[22]

As Oliphant knew from stinging experience, Rutherford regularly and vigorously espoused his belief that the atom was a sink for energy, rather than a reservoir. But in quieter moments, he also let slip his worries about whether his beloved nucleus might one day be mined for its rich treasures.

In the aftermath of one atom-splitting experiment, when asked where the research path led from that point, Rutherford replied wistfully: 'Who knows? We are entering no-man's land.'[23]

Yet he also had a canny idea of where such pursuits might lead. At a Royal Society Banquet in 1930, Rutherford approached Lord Hankey – then Secretary of the Committee of Imperial Defence – to discreetly canvass a topic the older professor thought was of some significance. Rutherford's unerring instincts told him the nuclear transformation experiments he was overseeing at the Cavendish, years before fission was a known concept, might one day prove to be important to Britain's defence. He was not sure precisely how, but he counselled Hankey that it might be worthwhile to 'keep an eye on the matter'.[24]

It was in recognition of these ambiguities, as much as an unwillingness to speak for his departed and much-missed friend and teacher, that Oliphant routinely parried the question about Rutherford's likely response to the bomb.

Oliphant would rightly assert that Rutherford played no direct part in the evolution of nuclear weapons; after all, his death came a year before the discovery of uranium fission. However, Oliphant also acknowledged that the atomic quest launched by that finding – then pursued at gathering speed to its horrific conclusion – had its roots firmly in Rutherford's Cavendish Laboratory.

He did not live to experience the excitement created by the discovery by Hahn and Strassmann, in 1938, of the fission process, or the beautiful work of Frisch and Meitner which established clearly that the uranium nucleus could indeed split into two parts when it absorbed a neutron.

If he had lived, he would have rejoiced in the subsequent triumphs of Lawrence and his colleagues in the Radiation Laboratory. But he would have regretted that his nuclear atom had become of such practical importance that the main motives for the financial support of such work, in all countries, became other than the advance of the knowledge of nature.[25]

Oliphant had been drawn to nuclear physics by the purity of Rutherford's vision. His lingering remorse over science's subversion was mitigated by relief that Rutherford remained eternally oblivious to the causes for which his nuclear insights were taken hostage.

'I am sure that he would agree that the preservation of no "ism", no way of life, no political system, all of which are ephemeral, could justify the manufacture or use of these diabolic weapons,' Oliphant would reason.[26]

'Many who worked with him and knew him well have tried to visualise his reaction to the terrible possibilities which the new things in science make probable ... Although we accept the fruits of the new spirit and of the new regard of the world for the scientific wizards whom it fears, at least we know that his view was right and ours is wrong. For him, our compromise would have been impossible.'[27]

'We couldn't have done anything else,' Oliphant would ultimately say, ruminating on the conflict between his beloved science and the ends for which it was used, 'but we have killed a beautiful subject.'[28]

EPILOGUE

Birmingham and Australia, 1945 to 2000

'You're a bloody fool!'

Mark Oliphant was pacing up and down the platform of London's Victoria railway station with fellow expatriate Howard Florey, a recently ordained Nobel laureate three years Oliphant's senior. 'You know if you leave this country you'll be committing scientific *hari kari*,' Florey thundered above the shunting engines.

Oliphant and his family had arrived at the station as it bustled with early summer traffic in July 1950, preparing for the short ride to Tilbury Docks. There, on the opposite bank of the Thames Estuary at Gravesend, James Smith Olifent and his kin had embarked for their new life in Australia in 1854. And now, almost a century later, the RMS *Orcades* – a replacement for the vessel of the same name sunk by Germany during the war – would convey Mark, Rosa and teenagers Michael and Vivian Oliphant on their third across-the-world voyage inside a decade.

Mark Oliphant was about to take his boldest ever leap of faith: signing on as a founding father of Canberra's Australian National University. Florey – who, along with historian and fellow expatriate Keith Hancock, had declined a similar offer – had travelled to London from Oxford, where he was Professor of Pathology. He had come not only to bid farewell to Oliphant, but also to offer some parting counsel to his friend, who he feared was courting professional disaster.

EPILOGUE

* * *

Had he sought personal reward, Mark Oliphant might easily have returned to America at war's end. He was offered several senior positions, among them a role with his scientific soulmate Ernest Lawrence at Berkeley. All of them carried, as Oliphant pointedly advised Sir Edward Appleton during their tussles over Tube Alloys, 'a much greater salary than any available in the UK'.[1]

But Oliphant returned from America in early 1945 with the aim of completing his long-stalled cyclotron project at Birmingham, and transforming it into an even more powerful machine than the ones Lawrence had devised, based on ideas he had formulated while with the Manhattan Project. His ambitious plans for a proton-synchrotron – a design of particle accelerator as yet unknown to the scientific world that relied less heavily on giant magnets – would overshadow the vast machinery at rival institutions, including Berkeley.

The initial price tag of £200,000 (almost £8 million today) for equipment alone seemed reasonable to Oliphant, given the scale of investments he had witnessed during the war. What he failed to recognise was the marked shift in thinking and circumstances, now the world was at peace. He quickly found Britain's post-war economic situation even more restrictive than that he had confronted on first pursuing his cyclotron idea a decade earlier.

Consequently when, in 1946, he was invited to London's Savoy Hotel to meet with Australia's prime minister, Ben Chifley, and hear of bold plans to establish a world-leading research university in the nation's capital city – but academic backwater – Canberra, the physicist's curiosity was roused. It sounded, after all, a similar proposition to that which had lured Rutherford to Montreal a half-century before: a move that had brought historic results.

In the spirit of his partnerships with Rutherford and then Lawrence, Oliphant sensed an affinity for the vision outlined by Chifley, the son of a blacksmith who had risen to the nation's top office through the ranks of the railways. Oliphant's loss of faith through Churchill's deceit was partially redeemed as he listened to Chifley expound his views on the importance of universities to the global post-war recovery effort.

'I began to see that politics was not always just a power game,' Oliphant later reflected. 'That there were people like Chifley who thought very deeply and sincerely about the problems of mankind. I suddenly knew I had met a politician with a profound feeling for humanity.'[2]

Despite his own misgivings, and the warning of Florey that 'what you'll find when you get there ... [is] a lot of promises, and a hole in the ground', Oliphant's need to grasp fresh projects led him to make the decision to return to the land of his birth. He left the still-unfinished cyclotron in the hands of Philip Moon, who also succeeded him as Birmingham's Poynting Professor of Physics.

With Peto packed up, the family sailed for the glorified country town turned national capital that – rather like the 'book colonies' from which Oliphant and Rutherford hailed – had been purpose-built from the foundation stone upwards.

Beset by teething problems, internecine squabbles and political interference from bureaucrats and rival universities, which foresaw precious funds being diverted to what they deemed an unnecessary project, the picture painted by Chifley of a brave new enterprise struggled to take shape. Oliphant spent more resources fighting administrative battles than pursuing meaningful science at the ANU. His primary dream of establishing the world's most powerful proton-synchrotron repeatedly stumbled into those holes of which Florey had warned, and was never realised in the form proposed. Reimagined and reconfigured as an ambitious homopolar generator, free of iron magnets, the huge machine was cruelly dubbed 'The White Oliphant'.

Almost from the time he arrived back in Australia, Mark Oliphant also faced personal trials. Barely a year after he had settled in Canberra, his mother Beatrice died aged eighty-three. Then, in 1963 – the year in which Mark's father Baron died – his now-adult son Michael began to develop numerous health problems. He underwent surgery in 1969, which found cancer in his stomach that had spread to his liver. He died in Melbourne in late January 1970, at the age of thirty-five and with his wife Monica more than eight months pregnant.

EPILOGUE

As Mark Oliphant's prior experience had taught him to do, he turned to his work to serve as grief's trusted outlet. By that time, his renown had grown to include a title.

In the first year of Queen Elizabeth II's reign in 1952, she agreed to appoint Oliphant a Knight Bachelor: an honour that he declined, just as he had in 1942. But he finally accepted a knighthood seven years later at the insistence of Prime Minister Robert Menzies, who convinced him it would aid the wider cause of scientific advancement in Australia: the same justification expressed by Rutherford when he was elevated to life peerage.

Oliphant was also chosen to lead Australia's delegation at the UN's first conference on Peaceful Uses of Atomic Energy held in Geneva in 1955. Two years later, he joined fellow scientists, including Max Born and Frederic Joliot-Curie, and several from the Soviet Union, at the inaugural Pugwash Conference on Science and World Affairs, which lobbied for the ban of nuclear weapons in war. It was a forum that Oliphant long considered among his most important post-war projects, along with establishing the Australian Academy of Science, an Antipodean offshoot of Britain's Royal Society, for which he received a charter from Queen Elizabeth II in 1954.

As the years passed, Oliphant's abhorrence for nuclear weapons evolved into a distaste towards atomic energy sourced from enriched uranium. Instead, he began to champion the potential of the universe's largest fusion generator – the sun – as the optimum means of meeting earth's energy needs.

After retiring from the ANU, he was named Governor of South Australia in 1971 and so became the first locally born appointee, as well as the first scientist to hold a role historically reserved for military men. He served a five-year term at Government House, where the immaculate grounds abutted Adelaide's former destitute asylum, once overseen by his great-grandfather, and the State Library where he had worked as an adolescent.

By the early 1980s, Rosa Oliphant was exhibiting symptoms of dementia, and following a serious fall in 1983 she was admitted to a nursing home in Adelaide. In January 1987, as her condition deteriorated, she was transferred to palliative care, where Mark, maintained a bedside vigil, just as Mary Rutherford had done to comfort her stricken husband fifty years earlier.

It was there, as Mark gently held her hand, from which the delicate wedding band he had fashioned had been removed to protect her fragile skin, that Rosa died calmly in her sleep.

Mark subsequently returned to Canberra, where he lived in a small flat at the rear of his daughter Vivian's home. Though his mind remained agile, he became increasingly physically frail and died on 14 July 2000, three months from his ninety-ninth birthday.

He had stipulated none of the ceremony or symbolism that had adorned Rutherford's funeral and internment at Westminster Abbey. Instead, Oliphant wanted the plainest of coffins, a perfunctory service attended only by immediate family, no flowers 'to wilt away', cremation as soon as practicable, and for his ashes to be scattered at Cleland Conservation Park in his cherished Adelaide Hills.

He wanted to be released to eternal rest within walking range of Mylor – as Cleland Park would certainly have been for the intrepid Oliphant boys, in the days when they trekked bush paths and forged their own tracks through the eucalyptus forests and flint-dry bracken, a world apart from the lush lawns and entrenched formality of Cambridge.

It was his home terrain, where the only scintillations witnessed were explosions of golden wattle against a screen of mottled green, transmuting to brown under summer's ferocity. Where the only laughter that split the stillness and echoed off ancient stones was the comforting peal of the kookaburra. Where the world's physical wonders endure, immutable and undisturbed, as they have for millennia.

Where he could be, forever, with nature.

* * *

'I regret it,' Oliphant reflected of his decision to return to Australia, more than thirty years after he left Britain. 'I think I was a fool to come back, from the intellectual and professional point of view. If you were to ask me … what part of my life have I enjoyed most, or what part of my life was most creative, then I'd have to say the Cambridge period.

'That was ... the happiest time of my life. Because of the whole spirit of the place and, of course, the attitude of Rutherford and his friendship. And the fact that one was discovering, every day, new things about nature.'[3]

Rather like those early revelations about the invisible bonds that connect the building blocks of the universe, the forces that united Ernest Rutherford and Mark Oliphant were improbable, yet irresistible.

Rutherford emerged from the volcanic plains and flax fields of rural New Zealand to not only shock the staid old world of European physics, but also to dictate its direction as expedition leader on the last great voyage of scientific discovery. Oliphant followed him – on the strength of a fleeting, wordless encounter in the colonial afterthought of Adelaide – to Cambridge University and the Cavendish Laboratory: the epicentre of scientific enterprise in an era that would reshape life ever after.

Their time together, from Oliphant's arrival until Rutherford's needlessly premature death, was comparatively brief. It spanned barely a decade, a period of global depression and radical upheaval. But from the field of nuclear physics that Rutherford so famously founded, and that Oliphant too pursued with almost fanatical zeal, came the innovations that define twenty-first-century life. Those advances can be traced back to the moment in 1917 when Rutherford's team successfully split the atom, and they vindicate his celebrated catchphrase that 'all science is either physics or stamp collecting'.[4]

'People show anger with science for developing nuclear power,' Oliphant reasoned to a popular weekly British news magazine in 1945. '"How soon", they ask, "before this terrible force annihilates us?" They should instead be asking, "How soon can it be made to serve us, end our drudgery, lift our loads?" We are already able to produce from a given weight of matter a million times more energy than has ever been obtained from it before.'[5]

It was not only its potential to deliver seemingly limitless resources of energy that ensured nuclear physics changed, irrevocably, the modern world. Without it, there would no radiotherapy in hospital cancer wards, nor the range of diagnostic imaging devices that can scan every part of the inner body without invasive surgery. The cathode tubes that delivered pictures in analogue television sets were, in essence, miniature particle accelerators. The digital flat screens that replaced them, and that are now the cornerstone of communication on mobile phones, computers and tablets, use technology sourced from the quantum theories spawned by Rutherford's work.

The worldwide web emerged from a small academic research tool devised by scientists working at the European Organization for Nuclear Research (CERN, the organisation that later developed the Large Hadron Collider) in Switzerland. Vital measurements of environmental and climactic changes from all parts of the globe are collated, around the clock, using accelerator mass spectrometry – by the modern descendants of tools integral to research at the Cavendish, and other physics laboratories that followed.

It was also physics that made possible nuclear weaponry, which renders all these advances potential hostage to a handful of political leaders. It's that final reality that came to haunt scientists, like Oliphant, who earnestly thought the development of such brutal weapons would rid the planet of global conflagrations. They enabled the bomb's birth in the naïve belief that, once the world had witnessed its terrifying power, no regime – not even one espousing Hitler's psychotic disregard for humanity – would risk initiating conflict, in the knowledge that reprisals would bring extinction.

The cessation of global warfare since Japan was bombed in 1945 might circumstantially suggest that Oliphant and his brethren of true believers have been vindicated. The reality, however, is that the weapon supposed to act as the ultimate deterrent has instead been wantonly brandished as the definitive threat. Seven decades have passed since the world's nuclear stockpile stood at zero after the destruction of Hiroshima and Nagasaki, and since Oliphant and Oppenheimer sought to seal a weapons pact at the United Nations. Subsequently eight other sovereign states have developed bomb capabilities. The global stockpile, which peaked at an estimated 70,000 warheads in the mid-1980s, currently stands around 15,000.[6]

Rather than safeguard the planet, the bomb has placed its very future at the whim of a handful of warmongers – some of whom are demonstrably more stable than others. As North Korea publicly

EPILOGUE

reaffirmed its aim to join that nuclear fraternity in 2017, United States President Donald Trump – commander-in-chief of the world's second-largest nuclear arsenal, behind Russia's – warned that any strike against American sovereign territory 'will be met with fire and fury like the world has never seen'.[7] Of this fire and fury we continue to live in fear.

None of this calamity loomed on that drizzly October morning in 1927 when Mark Oliphant was first introduced to Ernest Rutherford in the dark confines of the Cavendish Laboratory, two men from the far side of the world, neither knowing how that moment would bend the future. Yet it was the very commonality of their colonial pasts and their mutual devotion to science, at times to the detriment of those around them, that brought their relationship to bloom.

As twenty-something young men, they had both left behind outdoor boyhoods spent exploring untamed surrounds, mastering backyard engineering, and experiencing nature's raw beauty. Those experiences were tinged, on occasion, by tragic realities. But as well as being kindred souls, they were rare gems hewn from unique environments.

Rutherford, the Nobel laureate and British lord who would rest forever within the sacred heart of London's Westminster Abbey, was the world's most famous scientist when Oliphant arrived at Cambridge, justifiably daunted by the great man's reputation and aura. However, from that initial encounter, such an easy, sincere rapport developed that Oliphant became the master's master-craftsman, as well as his trusted laboratory partner, personal confidant and loyal companion. For all that Oliphant would achieve as a decisive and possibly divisive wartime figure, and as a public intellectual thereafter, it was his time with Rutherford that he held dearest. And most poignantly missed.

Their journey carried them from the distant Antipodes to the heart of matter, from the coal-cloaked valleys of Snowdonia to the golden age of physics. They were remarkable men who found each other at a most extraordinary time, and gave their gifts, without reservation, to their beloved science.

The world remains forever changed by their partnership.

ENDNOTES

Prologue
1. 'A Famous Scientist', *The Register* (Adelaide), 4 September 1925, p 11.
2. *Ibid.*
3. Newspaper clippings, undated and unattributed, Oliphant Collection, Rare Books and Special Collections, University of Adelaide, Series 24.
4. Stewart Cockburn and David Ellyard, *Oliphant*, Axiom Books, Adelaide, 1981, p 24.
5. *Ibid.*
6. Mark Oliphant, *Rutherford: Recollections of the Cambridge Days*, Elsevier, Amsterdam, 1972, p 18.
7. Robyn Williams, interview with Mark Oliphant, *The Science Show*, ABC Radio, Sydney, 1986.
8. Robin Hughes, interview with Mark Oliphant, *Australian Biography*, 20 January 1992, www.australianbiography.gov.au/subjects/oliphant/interview1.html.
9. *Ibid.*

1. Colonial Boys
1. 'South Australian Association for Emigration', *The Times* (London), 1 July 1834, p 4.
2. Philippa Mein Smith, *A Concise History of New Zealand*, Cambridge University Press, Cambridge and Melbourne, 2005, p 56.
3. R.M. Gibbs, *Under the Burning Sun*, Southern Heritage, Adelaide, 2013, p 229.
4. John Campbell, *Rutherford: Scientist Supreme*, AAS Publications, Christchurch, 1999, p 13.
5. Robin Hughes, *op cit*.
6. Jan Polkinghorne, *Mylor: Valley of Dreams*, Lutheran Publishing House, Adelaide, 1991, p 118.
7. Robin Hughes, *op cit*.
8. Mark Oliphant, speech at Ballarat College, December 1967, Oliphant Collection, Series 4.
9. Stewart Cockburn, interview with Keith Oliphant, Oliphant Collection, Series 26.
10. Campbell, *op cit*, p 13.
11. Robin Hughes, *op cit*.

2. The World Awaits
1. Cockburn, interview with Keith Oliphant, Oliphant Collection, Series 26.
2. Robin Hughes, *op cit*.

3. Cockburn, interview with Keith Oliphant, Oliphant Collection, Series 26.
4. Campbell, *op cit*, p 56.
5. Mark Oliphant, speech at Unley High School 75th jubilee dinner, 18 May 1985, Oliphant Collection, Series 4.
6. Robin Hughes, *op cit*.
7. Cockburn and Ellyard, *op cit*, p 20
8. *Ibid*.
9. Campbell, *op cit*, p 184.
10. Richard Reeves, *A Force of Nature: The Frontier Genius of Ernest Rutherford*, Atlas & Co, New York, 2008, p 29.
11. David Wilson, *Rutherford: Simple Genius*, MIT Press, Cambridge, Massachusetts, 1983, p 61.

3. 'Rabbit from the Antipodes'
1. Arthur Eve, *Rutherford: The Life and Letters of the Rt Hon. Lord Rutherford O.M.*, Cambridge University Press, Cambridge, 1939, p 15.
2. Wilson, *op cit*, p 64.
3. Hans C. Ohanian, *Einstein's Mistakes: The Human Failings of Genius*, Norton & Co, New York, 2008, p 25.
4. Malcolm Longair, *Maxwell's Enduring Legacy: A Scientific History of the Cavendish Laboratory*, Cambridge University Press, Cambridge, 2016, p 55.
5. Wilson, *op cit*, p 61.
6. Eve, *op cit*, p 14.
7. Reeves, *op cit*, p 34.
8. Herbert Childs, *An American Genius: The Life of Ernest Orlando Lawrence*, E.P. Dutton & Co, New York, 1968, p 210.
9. Eve, *op cit*, p 37.
10. Timothy Jorgensen, *Strange Glow: The Story of Radiation*, Princeton University Press, Princeton, 2016, p 26.
11. Diana Preston, *Before the Fall-Out: From Marie Curie to Hiroshima*, Corgi Books, London, 2005, p 54.
12. Reeves, *op cit*, p 38.
13. Eve, *op cit*, p 43.
14. *Ibid*, p 52.
15. Per Dahl, *Flash of the Cathode Rays: A History of J.J. Thomson's Electron*, Institute of Physics Publishing, Bristol, 1997, p 148.
16. Eve, *op cit*, p 50.
17. *Ibid*, p 55.

4. 'They'll Have Our Heads Off'
1. Thaddeus J. Trenn, *The Self-Splitting Atom: The History of the Rutherford–Soddy Collaboration*, Taylor & Francis, London, 1977, p 26.
2. Eve, *op cit*, p 77.
3. Trenn, *op cit*, p 26.
4. Reeves, *op cit*, p 48.
5. *Ibid*, pp 54, 64.
6. *Ibid*, p 64.
7. *Ibid*, p 68.
8. *Ibid*, p 51.
9. *Ibid*, p 39.

10. *Ibid*, p 55.
11. *Ibid*, p 76.
12. Robin McKown, *Giant of the Atom: Ernest Rutherford*, Julian Messner, New York, 1962, p 82.
13. Reeves, *op cit*, p 51.
14. Edward Andrade, *Rutherford and the Nature of the Atom*, Heinemann, London, 1965, p 73.
15. Spencer Weart, 'From the Nuclear Frying Pan into the Global Fire', *Bulletin of the Atomic Scientists*, Vol 48, No 5, June 1992, p 20.
16. Richard Rhodes, *The Making of the Atomic Bomb*, Simon & Schuster, New York, 1986, p 44.
17. Eve, *op cit*, p 291.

5. The Atom Smasher
1. Eve, *op cit*, p 164.
2. *Ibid*, p 157.
3. *Ibid*, p 158.
4. *Ibid*, p 183.
5. Reeves, *op cit*, p 71.
6. Eve, *op cit*, p 46.
7. *Ibid*, p 190.
8. *Ibid*, p 239.
9. Preston, *op cit*, p 66.
10. Mark Oliphant, 'Nuclear Physics and the Future', 37th Kelvin Lecture to the Institution of Electrical Engineers (UK), 25 April 1946, Special Collections, Cadbury Research Library, University of Birmingham, 30F/3.
11. Arthur Eddington, *The Nature of the Physical World: Gifford Lectures*, Cambridge University Press, Cambridge 1928, p 1.
12. John Hendry (ed), *Cambridge Physics in the Thirties*, Adam Hilger, Bristol, 1984, p 138.
13. *Ibid*, p 87.
14. Reeves, *op cit*, p 84.
15. Eve, *op cit*, p 224.
16. Wilson, *op cit*, p 340.
17. Eve, *op cit*, p 233.
18. Ernest Rutherford, 'Henry Gwyn Jeffreys Moseley', *Nature*, Vol 96, No 2393, 9 September 1915, p 33.
19. Wilson, *op cit*, p 345.
20. Campbell, *op cit*, p 445.
21. Egon Larsen, *The Cavendish Laboratory: Nursery of Genius*, Edmund Ward, London, 1962, p 52.
22. Eve, *op cit*, p 264.
23. Rhodes, *op cit*, p 137.

6. A Benevolent Lord
1. Eve, *op cit*, p 269.
2. *Ibid*.
3. *Ibid*.
4. *Ibid*.
5. Longair, *op cit*, p 184.

6. Ferenc Morton Szasz, *British Scientists and the Manhattan Project*, Macmillan Academic & Professional, Basingstoke, 1992, p 77.
7. Larsen, *op cit*, p 63.
8. Michael Hiltzik, *Big Science*, Simon & Schuster, New York, 2015, p 226.
9. *Ibid*, p 273.
10. *Ibid*, p 281.
11. *Ibid*, p 304.
12. Eve, *op cit*, p 310.
13. *Ibid*, p 311.
14. David Cassidy, *J. Robert Oppenheimer and the American Century*, Pi Press, New York, 2005, p 51.
15. *Ibid*, p 95.
16. Robert Ditchburn, 'Reminiscences', in Rajkumari Williamson (ed), *The Making of Physicists*, Adam Hilger, Bristol, 1987, p 19.
17. Cassidy, *op cit*, p 95.
18. Frederick Mann, *Lord Rutherford on the Golf Course*, Cambridge University Press, Cambridge, 1976, p 25.
19. Eve, *op cit*, p 317.

7. 'A Rare Quality of Mind'

1. Mick Joffe, 'Sir Mark Oliphant: Reluctant Builder of the Atom Bomb', 1996, Mick Joffe Caricatures, www.mickjoffe.com/Sir_Mark_Oliphant.
2. Robin Hughes, *op cit*.
3. Williams, *op cit*.
4. Cockburn and Ellyard, *op cit*, p 24.
5. *Ibid*.
6. *Ibid*.
7. *Ibid*, p 29.
8. Ann Moyal, *Portraits in Science*, National Library of Australia, Canberra, 1994, p 22.
9. Cockburn and Ellyard, *op cit*, p 29.
10. Kerr Grant, 'Reference for Mark Oliphant', 3 July 1928, Oliphant Collection, Series 26.
11. Roy Burdon, 'Reference for Mark Oliphant', 13 May 1927, Oliphant Collection, Series 26.
12. Cockburn and Ellyard, *op cit*, p 30.

8. String and Sealing Wax

1. Robin Hughes, *op cit*.
2. *Ibid*.
3. Oliphant, *Rutherford*, p 19.
4. *Ibid*, p 21.
5. Mark Oliphant, speech to the Australian Labor Party, Canberra, 29 November 1968, Oliphant Collection, Series 4.
6. Oliphant, *Rutherford*, p 20.
7. *Ibid*, p 21.
8. Philip Moon, letter to Mark Oliphant, 18 May 1970, Oliphant Collection, Series 2.
9. Cockburn and Ellyard, *op cit*, p 37.
10. Oliphant, *Rutherford*, p 21.

11. David Ellyard, interview with Elizabeth Cockcroft, Oliphant Collection, Series 26.
12. Cockburn, interview with Rosa Oliphant, Oliphant Collection, Series 26.
13. Hendry, *op cit*, p 86.
14. Oliphant, speech to the Australian Labor Party.
15. Campbell, *op cit*, p 466.
16. Hiltzik, *op cit*, p 120.

9. A Meeting of Minds
1. John Chadwick, 'Foreword', in Oliphant, *Rutherford*, p xi.
2. Oliphant, *Rutherford*, p 130.
3. Campbell, *op cit*, p 446.
4. Oliphant, *Rutherford*, p 126.
5. *Ibid*, p 124.
6. Mark Oliphant, speech to the Physical Society (UK), 7 October 1946, Special Collections, University of Birmingham, 30F/3.
7. Lawrence Badash, 'Nagaoka to Rutherford, 22 February 1911', *Physics Today*, Vol 20, No 4, April 1967, p 55.
8. Campbell, *op cit*, p 450.
9. Oliphant, *Rutherford*, p 145.
10. Charles Wynn-Williams, letter to Mark Oliphant, 24 July 1970, Oliphant Collection, Series 2.
11. Mark Oliphant, speech, untitled and undated, delivery location unknown, Oliphant Collection, Series 4.
12. Reeves, *op cit*, p 109.
13. Mark Oliphant, 'The Significance of Rutherford Today', speech at the University of Kent, 30 October 1971, Oliphant Collection, Series 4.
14. Oliphant, *Rutherford*, p 24.
15. Evelyn Shaw, letter to Mark Oliphant, 29 November 1929, Oliphant Collection, Series 2.
16. Campbell, *op cit*, p 467.
17. *Ibid*, p 420.
18. *Ibid*, p 422.
19. *Ibid*.
20. *Ibid*, p 423.
21. Campbell, *op cit*, p 422
22. Cockburn and Ellyard, *op cit*, p 44.
23. Campbell, *op cit*, p 424.
24. *Ibid*
25. *Ibid*, p 425.
26. *Ibid*, p 426.
27. Oliphant, *Rutherford*, p 125.

10. The Golden Year
1. Longair, *op cit*, p 213.
2. Reeves, *op cit*, p 115.
3. Oliphant, *Rutherford*, p 74.
4. *Ibid*, p 75.
5. Peter Watson, *A Terrible Beauty: A History of the People and Ideas that Shaped the Modern Mind*, Weidenfeld & Nicolson, London, 2000, p 264.
6. Oliphant, *Rutherford*, p 76.

7. Reeves, *op cit*, p 116.
8. Rhodes, *The Making of the Atomic Bomb*, p 209.
9. Mark Oliphant, letter to A.E. Kempton, 19 May 1970, Oliphant Collection, Series 2.
10. John Fremlin, letter to Mark Oliphant, 21 May 1970, Oliphant Collection, Series 2.
11. National Aeronautics and Space Administration (NASA) Science News, *Human Voltage*, published 18 June 1999, https://science.nasa.gov/science-news/science-at-nasa/1999/essd18jun99_1.
12. Larsen, *op cit*, p 72.
13. Oliphant, *Rutherford*, p 86.
14. Rhodes, *The Making of the Atomic Bomb*, p 165.
15. John Chadwick, 'Foreword', in Oliphant, *Rutherford*, p x.

11. Fusion
1. Oliphant, *Rutherford*, p 106.
2. Robin Hughes, *op cit*.
3. Oliphant, *Rutherford*, p 110.
4. John Hughes, 'Proposals in Support of the Candidature of Professor Sir Mark Oliphant for the 1976 Nobel Prize in Physics', Oliphant Collection, Series 26.
5. Richard Rhodes, *Dark Sun: The Making of the Hydrogen Bomb*, Simon & Schuster, New York, 1995, p 247.
6. Oliphant, *Rutherford*, p 111.
7. Cockburn, interview with Rosa Oliphant, Oliphant Collection, Series 26.
8. Cockburn and Ellyard, *op cit*, p 51.
9. Robin Hughes, *op cit*.

12. Tyranny's Dark Clouds
1. Rhodes, *The Making of the Atomic Bomb*, p 27.
2. Oliphant, *Rutherford*, p 141.
3. Campbell, *op cit*, p 488.
4. Mark Oliphant, 'Rutherford Memorial Lecture to Royal Society 1955', Oliphant Collection, Series 4.
5. Oliphant, *Rutherford*, p 141.
6. *Ibid*, p 57.
7. *Ibid*, p 59.
8. *Ibid*.

13. The Crown Begins to Slip
1. Mark Oliphant, 'The Two Ernests', undated, p 11, Oliphant Collection, Series 26.
2. Longair, *op cit*, p 225.
3. Oliphant, *Rutherford*, p 40.
4. *Ibid*.
5. *Ibid*, p 138.
6. *Ibid*, p 107.
7. Rhodes, *The Making of the Atomic Bomb*, p 159.
8. Campbell, *op cit*, p 431.
9. Oliphant, *Rutherford*, p 132.
10. *Ibid*, p 128.
11. Campbell, *op cit*, p 420.

12. Cockburn and Ellyard, *op cit*, p 61.
13. *Ibid*, p 62.
14. Longair, *op cit*, p 191.
15. Oliphant, 'The Two Ernests', p 17, Oliphant Collection, Series 26.
16. Mark Oliphant, letter to James Chadwick, 20 April 1967, Oliphant Collection, Series 26.
17. *Ibid*.
18. Oliphant, speech to the Physical Society (UK), Special Collection, University of Birmingham.
19. Cockburn and Ellyard, *op cit*, p 61.
20. Hendry, *op cit*, p 131.
21. Stewart Cockburn, interview with Mark Oliphant, June 1980, Oliphant Collection, Series 26.
22. Robin Hughes, *op cit*.
23. Cockburn and Ellyard, *op cit*, p 68.
24. Longair, *op cit*, p 281.

14. 'Requiem Aeternam'
1. Oliphant, *Rutherford*, p 155.
2. Campbell, *op cit*, p 472.
3. Niels Bohr, 'The Right Hon. Lord Rutherford of Nelson, O.M., F.R.S', *Nature*, Vol 140, No 3548, 30 October 1937, p 752.
4. Campbell, *op cit*, p 474.
5. Mark Oliphant, letter to Philip Dee, 15 July 1970, Oliphant Collection, Series 2.
6. Bohr, *op cit*, p 754.
7. Larsen, *op cit*, p 64.
8. Mark Oliphant, 'Some Personal Recollections of Rutherford, the Man', *Notes and Records of the Royal Society*, Vol 27, No 1, August 1972, p 20.
9. Mark Oliphant, letter to A.P. Rowe, undated, Oliphant Collection, Series 26.
10. Oliphant, *Rutherford*, p 142.
11. *Ibid*, p 157.
12. Reeves, *op cit*, p 169.
13. Cockburn and Ellyard, *op cit*, p 65.
14. *Ibid*.
15. Mary Rutherford, letter to Mark Oliphant, October 1937, Oliphant Collection, Series 24.

15. 'A Show of My Own'
1. Mark Oliphant, letter to Niels Bohr, 13 December 1937, Special Collections, University of Birmingham, 30/F3.
2. 'A British Testament of Faith – On Leaving Birmingham', *Birmingham Mail*, 30 June 1950, Special Collections, University of Birmingham, Scrapbooks 22X/1A G9.
3. 'University Asks for £60,000', *Birmingham Post*, 7 April 1938, Special Collections, University of Birmingham.
4. Mark Oliphant, 'The Genesis of the Nuffield Cyclotron', Department of Physics, 1967, p 2, Special Collections, University of Birmingham, 22X/8.
5. *Ibid*, p 4.
6. *Ibid*, p 5.
7. '£60,000 Gift in Five Words', *Birmingham Gazette*, 30 June 1938, Special Collections, University of Birmingham, 22X/8.

8. 'The Scientist Works for Industry', *Birmingham Gazette*, 6 July 1938, Special Collections, University of Birmingham, 22X/8.
9. Hiltzik, *op cit*, p 114.
10. Oliphant, 'The Two Ernests', p 12, Oliphant Collection, Series 26.
11. *Ibid*, p 1.
12. John Heilbron and Robert Seidel, *Lawrence and His Laboratory*, University of California Press, Los Angeles, 1989, p 350.
13. 'Black Magic at the Varsity', *Evening Despatch* (Birmingham), 27 April 1939, Special Collections, University of Birmingham, 22X/1A G9.
14. 'Students Make an Atom Splitter that May Create Energy', *Newcastle Journal and North Mail*, 1 May 1939, Special Collections, University of Birmingham, 22X/8.
15. Rhodes, *The Making of the Atomic Bomb*, p 259.
16. David Gregory, 'On the Right Side of Wrong', interview with Otto Frisch, BBC Midlands, Birmingham, 17 February 2000.
17. Ernest Rutherford, 'The Electrical Structure of Matter', *Annual Report of the Board of Regents of the Smithsonian Institution 1924*, Publication 2795, Washington, DC, 1925, p 180.
18. Rhodes, *The Making of the Atomic Bomb*, p 261.
19. Hiltzik, *op cit*, p 214.
20. Rhodes, *The Making of the Atomic Bomb*, p 266.
21. 'Scientists Make an Amazing Discovery', *Sunday Express* (London), 30 April 1939, Special Collections, University of Birmingham, 22X/1A G9.
22. Mark Oliphant, letter to Niels Bohr, 30 May 1939, Special Collections, University of Birmingham, 30/F3.

16. The Decisive Difference
1. Cockburn and Ellyard, *op cit*, p 82.
2. Neville Chamberlain, 'Declaration of War', 30 September 1939, BBC Archives www.bbc.co.uk/archive/ww2outbreak/7957.shtml?page=txt.
3. Raymond Priestley, diary, September 1939, Special Collections, University of Birmingham, XUS38 2/2.
4. Ted Nield, 'Beyond the Call of Duty', *New Scientist*, Vol 123, No 1681, 9 September 1989, p 77.
5. Mark Oliphant, 'Sir Mark Oliphant', in Peter M. Rolph (ed.), *Fifty Years of the Cavity Magnetron: Proceedings of a One-Day Symposium, 21 February 1990*, School of Physics and Space Research, University of Birmingham, Birmingham, 1990, p 10.
6. Ronald Clark, *Birth of the Bomb*, Phoenix House, London, 1961, p 87.
7. *Ibid*.
8. David Ellyard, interview with John Randall, Oliphant Collection, *op cit*, Series 26.
9. Cockburn and Ellyard, *op cit*, p 183.
10. Oliphant in Rolph (ed.), *Fifty Years of the Cavity Magnetron*, p 10.
11. *Ibid*.
12. Albert Rowe, *One Story of Radar*, Cambridge University Press, Cambridge, 1948, p 35.
13. James Phinney Baxter, *Scientists Against Time*, Little, Brown & Co, Boston, 1946, p 142.
14. Cockburn, interview with Rosa Oliphant, Oliphant Collection, Series 26.

17. 'Shouldn't Someone Know About This?'

1. Adolf Hitler, Danzig speech, 19 September 1939, in *Adolf Hitler: Collection of Speeches 1922–1945*, archive.org/details/AdolfHitlerCollectionOfSpeeches 19221945, p 636.
2. Hiltzik, *op cit*, p 222.
3. Cockburn and Ellyard, *op cit*, p 82.
4. Niels Bohr, letter to Mark Oliphant, 27 May 1939, Special Collections, University of Birmingham, 30/F3.
5. Oliphant, letter to Niels Bohr, 30 May 1939, *op cit*.
6. Otto Frisch, *What Little I Remember*, Cambridge University Press, Cambridge, 1979, p 120.
7. *Ibid*.
8. Rudolf Peierls, *Bird of Passage: Recollections of a Physicist*, Princeton University Press, Princeton, 1985, p 127.
9. *Ibid*, p 133.
10. Rudolf Peierls, *Atomic Histories*, American Institute of Physics, 1997, p 160.
11. Frisch, *What Little I Remember*, p 123.
12. Ronald Clark, *Tizard*, Methuen & Co, London, 1965, p 214.
13. Frisch, *What Little I Remember*, p 125.
14. Peierls, *Bird of Passage*, p 157.
15. *Ibid*.
16. Frisch, *What Little I Remember*, p 126.
17. *Ibid*.
18. Peierls, *Bird of Passage*, p 154.
19. *Ibid*, p 155.
20. Otto Frisch and Rudolf Peierls, 'Frisch–Peierls Memorandum, March 1940', Atomic Archive, www.atomicarchive.com/Docs/Begin/FrischPeierls.shtml.
21. Clark, *Tizard*, p 215.
22. *Ibid*, p 216.
23. *Ibid*.
24. Frisch, *What Little I Remember*, p 126.
25. 'One Man and His Bomb', *Independent*, 9 July 1995, www.independent.co.uk/arts-entertainment/one-man-and-his-bomb-1590614.html
26. Clark, *Tizard*, p 218.

18. MAUD

1. Clark, *Tizard*, p 218.
2. Rhodes, *The Making of the Atomic Bomb*, p 296.
3. Henry Tizard, letter to Wing Commander William Elliott, 4 May 1940, National Archives, London, Series AB 1/222.
4. Frisch, *What Little I Remember*, p 128.
5. Guy Hartcup and Thomas Allibone, *Cockcroft and the Atom*, Adam Hilger, Bristol, 1984, p 121.
6. Clark, *Tizard*, p 221.
7. Oliphant in Rolph (ed.), *Fifty Years of the Cavity Magnetron*, p 10.
8. *Ibid*.
9. David Nichols (ed), *Ernie's War: The Best of Ernie Pyle's World War II Dispatches*, Simon & Schuster, New York, 1986, p 42.
10. Cockburn and Ellyard, *op cit*, p 82.
11. Mark Oliphant, written recollections (undated), Oliphant Collection, Series 20.

12. 'Report by the MAUD Committee on the Use of Uranium for a Bomb', National Archives, Series AB 1/238.
13. Rhodes, *The Making of the Atomic Bomb*, p 343.
14. 'Report by the MAUD Committee on the Use of Uranium for a Bomb', National Archives.
15. *Ibid*.
16. Rhodes, *The Making of the Atomic Bomb*, p 372.
17. *Ibid*.
18. Cockburn and Ellyard, *op cit*, p 109.
19. Graham Farmelo, *Churchill's Bomb*, Faber & Faber, London, 2013, p 200.
20. Clark, *Tizard*, p 298.
21. Silvan Schweber, *Einstein and Oppenheimer: The Meaning of Genius*, Harvard University Press, Cambridge, Massachusetts, 2008, p 45.
22. Jeremy Bernstein, *Hitler's Uranium Club: The Secret Recordings at Farm Hall*, American Institute of Physics, Woodbury, 1996, p 14.
23. Archibald Hill, 'Uranium-235', 16 May 1940, National Archives London, Series AB 1/222.

19. 'Meddling Foreigner'
1. Mark Oliphant, letter to George Thomson, 9 August 1941, National Archives, London, Series AB 15/6077.
2. Mark Oliphant, 'The Beginning: Chadwick and the Neutron', *Bulletin of the Atomic Scientists*, Vol 38, No 10, December 1982, p 17.
3. Jeremy Bernstein, 'A Memorandum that Changed the World', *American Journal of Physics*, Vol 79, No 5, May 2011, p 446.
4. Williams, *op cit*.
5. Nuel Pharr Davis, *Lawrence and Oppenheimer*, Jonathan Cape, London, 1969, p 112.
6. William Coolidge, letter to Frank Jewett, 13 September 1941, Oliphant Collection, Series 11.
7. Davis, *op cit*, p 113.
8. Hiltzik, *op cit*, p 15.
9. Rhodes, *The Making of the Atomic Bomb*, p 373.
10. Oliphant, 'The Two Ernests', p 52, Oliphant Collection, Series 26.
11. Peter Michelmore, *The Swift Years: The Robert Oppenheimer Story*, Dodd, Mead & Co, New York, 1969, p 66.
12. Mark Oliphant, 'MAUD: Notes on Conversation with E.O. Lawrence in Berkeley', 25 September 1941, National Archives, London, Series AB 1/709.
13. Cockburn and Ellyard, *op cit*, p 108.
14. Szasz, *op cit*, p 5.
15. Spencer Weart and Gertrud Weiss Szilard, *Leo Szilard: His Version of the Facts*, MIT Press, Cambridge, 1978, excerpt in *The Bulletin* (Sydney), April 1979, p 29.
16. Cockburn and Ellyard, *op cit*, p 108.
17. Williams, *op cit*.
18. Gregg Herken, *Brotherhood of the Bomb: The Tangled Lives of Robert Oppenheimer, Ernest Lawrence and Edward Teller*, Henry Holt & Co, New York, 2002, p 40.
19. Davis, *op cit*, p 114.
20. Franklin Roosevelt, letter to Winston Churchill, 11 October 1941, National Archives, London, Series PREM 3/139/8A.
21. Raymond Priestley, diary, 6–14 October 1941, Special Collections, University of Birmingham, XUS38 2/4.

20. A Misguided Mission
1. Cockburn, interview with Rosa Oliphant, Oliphant Collection, Series 26.
2. Cockburn and Ellyard, *op cit*, p 113.
3. Raymond Priestley, diary, 21–28 February 1942, Special Collections, University of Birmingham, XUS38 2/4.
4. Cockburn and Ellyard, *op cit*, p 92.
5. Cockburn, interview with Rosa Oliphant, Oliphant Collection, Series 26.
6. Robert Nimmo, draft of telegram to Mark Oliphant per David Rivett, Special Collections, University of Birmingham, 30F/3.
7. Mark Oliphant, telegram to Robert Nimmo, 3 June 1942, Special Collections, University of Birmingham, 30F/3.
8. J.M. Martin, letter to Robert Nimmo, 4 June 1942, Special Collections, University of Birmingham, 30F/3.
9. Cockburn and Ellyard, *op cit*, p 111.
10. Raymond Priestley, diary, 20 August 1945, Special Collections, University of Birmingham, XUS38 2/6.
11. Cockburn and Ellyard, *op cit*, p 112.
12. Szasz, *op cit*, p 6.
13. *Ibid*, p 12.

21. Manhattan
1. Oliphant, 'The Beginning: Chadwick and the Neutron', *op cit*, p 18.
2. Kenneth Nichols, 'The Road to Trinity', in Cynthia Kelly (ed), *The Manhattan Project*, Black Dog & Leventhal Publishers, New York, 2009, p 121.
3. Robert Serber, *Peace and War: Reminiscences of a Life on the Frontiers of Science*, Columbia University Press, New York, 1998, p 72.
4. Szasz, *op cit*, p xvii.
5. Robert Norris, *Racing for the Bomb: The True Story of General Leslie R. Groves*, Steerforth Press, South Royalton, Vermont, 2002, p 528.
6. Cockburn and Ellyard, *op cit*, p 112.
7. Mark Oliphant, letter to General Leslie Groves, 27 November 1943, Oliphant Collection, Series 26.
8. *Ibid*.
9. Mark Oliphant, letter to Ernest Lawrence, 16 February 1944, Oliphant Collection, Series 26.
10. Rhodes, *The Making of the Atomic Bomb*, p 373.
11. James Chadwick, letter to General Leslie Groves, 23 January 1945, Oliphant Collection, Series 26.
12. Oliphant, 'The Beginning: Chadwick and the Neutron', *op cit*, p 18.
13. Raymond Priestley, diary, 11–17 December 1944, Special Collections, University of Birmingham, XUS38 2/6.
14. Letter from Mark Oliphant to Ernest Lawrence, 16 March 1945, quoted in Cockburn and Ellyard, *op cit*, p 122.
15. Mark Oliphant, letter to General Leslie Groves, 3 July 1945, Oliphant Collection, Series 26.

22. 'Death, the Shatterer of Worlds'
1. Daniel Ellsberg, *The Doomsday Machine: Confessions of a Nuclear War Planner*, Bloomsbury, New York, 2017, p 275.
2. Rhodes, *The Making of the Atomic Bomb*, p 419.

3. Serber, *op cit*, p 91.
4. Leslie Groves, 'Memorandum to Secretary of War: Subject – The Test', 18 July 1945, Oliphant Collection, Series 26.
5. *Ibid*.
6. *Ibid*.
7. Robin Hughes, *op cit*.
8. Winston Churchill, *Triumph and Tragedy*, Houghton Mifflin, Boston, 1953, p 638.
9. Rhodes, *The Making of the Atomic Bomb*, p 699.
10. *Ibid*.

23. 'We Have Killed a Beautiful Subject'
1. 'The First Real Holiday', *The Times* (London), 6 August 1945, p 2.
2. Wallace Akers, telegram to Tube Alloys scientists, 4 August 1945, National Archives, London, Series AB 1/53.
3. 'The First Atomic Bomb Hits Japan', *The Times* (London), 7 August 1945, p 4.
4. Richard Kisch, 'Mark Oliphant Wages War on Bomb Secrecy', *The News* (Adelaide), 17 November 1945, p 2.
5. 'The First Atomic Bomb Hits Japan', *The Times* (London), *op cit*.
6. *Ibid*.
7. Hugh Warren, letter to Wallace Akers, 8 August 1945, National Archives, London, Series AB 1/53.
8. Stewart Cockburn, interview with David Robertson, Oliphant Collection, Series 26.
9. Cockburn, interview with Mark Oliphant, Oliphant Collection, Series 26.
10. Serber, *op cit*, p 91.
11. Mark Oliphant, article for *Sydney Morning Herald*, 19 August 1979, Oliphant Collection, Series 20.
12. Williams, *op cit*.
13. Cockburn and Ellyard, *op cit*, p xiii.
14. Joffe, *op cit*.
15. Cockburn and Ellyard, *op cit*, p 126.
16. Anna Rothe (ed), *Current Biography: Who's New and Why*, H.H. Wilson Co, New York, 1951, p 468.
17. Cockburn and Ellyard, *op cit*, p 134.
18. Margaret Gowing, *Independence and Deterrence: Britain and Atomic Energy 1945–1952*, Macmillan Press, London, 1974, p 323.
19. Robin Hughes, *op cit*.
20. Mark Oliphant, interviewed on *The Scientists* radio program, 16 September (year not recorded), Oliphant Collection, Series 26.
21. Oliphant, 'The Significance of Rutherford Today', Oliphant Collection, Series 4.
22. *Ibid*.
23. Larsen, *op cit*, p 73.
24. Clark, *Birth of the Bomb*, p 153.
25. Oliphant, 'The Two Ernests', pp 51–52, Oliphant Collection, Series 26.
26. Oliphant, 'The Significance of Rutherford Today', Oliphant Collection, Series 4.
27. Oliphant, speech to the Physical Society (UK), Special Collections, University of Birmingham.
28. C.P. Snow, *The Physicists*, House of Stratus, Looe, Cornwall, 2001, p 2.

Epilogue
1. Mark Oliphant, letter to Edward Appleton, 14 February 1945, National Archives, London, Series AB 1/690.
2. Cockburn and Ellyard, *op cit*, p 146.
3. Robin Hughes, *op cit*.
4. Andrade, *op cit*, p 70.
5. Cockburn and Ellyard, *op cit*, p 130.
6. Hans M. Kristensen and Robert S. Norris, 'Status of World Nuclear Forces', Federation of American Scientists, June 2018, fas.org/issues/nuclear-weapons/status-world-nuclear-forces.
7. Peter Baker and Choe Sang-Hun, 'Trump Threatens "Fire and Fury" Against North Korea if it Endangers US', *New York Times*, 8 August 2017, p 1.

BIBLIOGRAPHY

Collections
Oliphant Papers, Rare Books and Special Collections, University of Adelaide
Series AB, PREM, National Archives, London
Somerville Oral History Collection, State Library of South Australia
Special Collections, Cadbury Research Library, University of Birmingham

Interviews
David Gregory, 'On the Right Side of Wrong', interview with Otto Frisch, BBC Midlands, Birmingham, 17 February 2000
Robin Hughes, interview with Mark Oliphant, *Australian Biography*, 20 January 1992, www.australianbiography.gov.au/subjects/oliphant/interview1.html
Mick Joffe, 'Sir Mark Oliphant: Reluctant Builder of the Atom Bomb', 1996, Mick Joffe Caricatures, www.mickjoffe.com/Sir_Mark_Oliphant
Andrew Ramsey, interview with Monica Oliphant, Adelaide, November 2017
Robyn Williams, interview with Mark Oliphant, *The Science Show*, ABC Radio, Sydney, 1986

Newspapers and Magazines
The Advertiser (Adelaide), trove.nla.gov.au
Birmingham Gazette
Birmingham Mail
Birmingham Post
The Bulletin (Sydney), trove.nla.gov.au
Evening Despatch (Birmingham)

Independent (London)
New York Times
Newcastle Journal and North Mail
The News (Adelaide), trove.nla.gov.au
The Register (Adelaide), trove.nla.gov.au
Sunday Express (London)
Sydney Morning Herald, trove.nla.gov.au
The Times (London)
Washington Post

Books, Journal Articles and Webpages
Edward Andrade, *Rutherford and the Nature of the Atom*, Heinemann, London, 1965
Lawrence Badash, 'Nagaoka to Rutherford, 22 February 1911', *Physics Today*, Vol 20, No 4, April 1967, pp 55–60
James Phinney Baxter, *Scientists Against Time*, Little, Brown & Co, Boston, 1946
Jeremy Bernstein, *Hitler's Uranium Club: The Secret Recordings at Farm Hall*, American Institute of Physics, Woodbury, New York, 1996
—— 'A Memorandum that Changed the World', *American Journal of Physics*, Vol 79, No 5, May 2011, pp 440–46
Kai Bird and Martin J. Sherwin, *American Prometheus: The Triumph and Tragedy of J. Robert Oppenheimer*, Alfred A. Knopf, New York, 2005
J.B. Birks (ed), *Rutherford at Manchester*, Heywood & Co, London, 1962
Brebis Bleaney, 'Sir Mark (Marcus Laurence Elwin) Oliphant A.C., K.B.E.', *Biographical Memoirs of Fellows of the Royal Society*, No 47, 1 November 2001, pp 383–93
Niels Bohr, 'The Right Hon. Lord Rutherford of Nelson, O.M., F.R.S', *Nature*, Vol 140, No 3548, 30 October 1937, pp 752–53
John Campbell, *Rutherford's Ancestors*, AAS Publications, Christchurch, 1996
—— *Rutherford: Scientist Supreme*, AAS Publications, Christchurch, 1999
David Cassidy, *J. Robert Oppenheimer and the American Century*, Pi Press, New York, 2005
Brian Cathcart, *The Fly in the Cathedral: How a Group of Cambridge Scientists Won the International Race to Split the Atom*, Viking, London, 2004

Neville Chamberlain, 'Declaration of War', 30 September 1939, BBC Archives www.bbc.co.uk/archive/ww2outbreak/7957.shtml?page=txt
Herbert Childs, *An American Genius: The Life of Ernest Orlando Lawrence*, E.P. Dutton & Co, New York, 1968
Winston Churchill, *Triumph and Tragedy*, Houghton Mifflin, Boston, 1953
Ronald Clark, *Birth of the Bomb*, Phoenix House, London, 1961
—— *Tizard*, Methuen & Co, London, 1965
Stewart Cockburn and David Ellyard, *Oliphant*, Axiom Books, Adelaide, 1981
Arthur Holly Compton, *Atomic Quest: A Personal Narrative*, Oxford University Press, London, 1956
J.G. Crowther, *British Scientists of the Twentieth Century*, Routledge & Kegan Paul, London, 1952
Marcus Cunliffe, 'America at the Great Exhibition of 1851', *American Quarterly*, Vol 3, No 2, Summer 1951, pp 115–26
Per Dahl, *Flash of the Cathode Rays: A History of J.J. Thomson's Electron*, Institute of Physics Publishing, Bristol, 1997
Nuel Pharr Davis, *Lawrence and Oppenheimer*, Jonathan Cape, London, 1969
Arthur Eddington, *The Nature of the Physical World: Gifford Lectures*, Cambridge University Press, Cambridge, 1928
Daniel Ellsberg, *The Doomsday Machine: Confessions of a Nuclear War Planner*, Bloomsbury, New York, 2017
Arthur Eve, *Rutherford: The Life and Letters of the Rt Hon. Lord Rutherford O.M.*, Cambridge University Press, Cambridge, 1939
Graham Farmelo, *Churchill's Bomb*, Faber & Faber, London, 2013
Otto Frisch, *What Little I Remember*, Cambridge University Press, Cambridge, 1979
Otto Frisch and Rudolf Peierls, 'Frisch–Peierls Memorandum, March 1940', Atomic Archive, www.atomicarchive.com/Docs/Begin/FrischPeierls.shtml
R.M. Gibbs, *Under the Burning Sun*, Southern Heritage, Adelaide, 2013
Margaret Gowing, *Britain and Atomic Energy 1939–1945*, Macmillan Press, London, 1964
—— *Independence and Deterrence: Britain and Atomic Energy 1945–1952*, Macmillan Press, London, 1974
Leslie R. Groves, *Now It Can Be Told: The Story of the Manhattan Project*, André Deutsch, London, 1963

Paul Ham, *Hiroshima Nagasaki*, HarperCollins, Sydney, 2011

Guy Hartcup and Thomas Allibone, *Cockcroft and the Atom*, Adam Hilger, Bristol, 1984

John Heilbron and Robert Seidel, *Lawrence and His Laboratory*, University of California Press, Los Angeles, 1989

John Hendry (ed), *Cambridge Physics in the Thirties*, Adam Hilger, Bristol, 1984

Gregg Herken, *Brotherhood of the Bomb: The Tangled Lives of Robert Oppenheimer, Ernest Lawrence and Edward Teller*, Henry Holt & Co, New York, 2002

James G. Hershberg, *James B. Conant: Harvard to Hiroshima – The Making of the Nuclear Age*, Alfred A. Knopf, New York, 1993

Richard G. Hewlett and Oscar E. Anderson Jr, *The New World 1939–1946: A History of the United States Atomic Energy Commission*, Pennsylvania State University, Pennsylvania, 1962

Michael Hiltzik, *Big Science*, Simon & Schuster, New York, 2015

Adolf Hitler, *Adolf Hitler: Collection of Speeches 1922–1945*, archive.org/details/AdolfHitlerCollectionOfSpeeches19221945

Timothy Jorgensen, *Strange Glow: The Story of Radiation*, Princeton University Press, Princeton, 2016

Cynthia Kelly (ed), *The Manhattan Project*, Black Dog & Leventhal Publishers, New York, 2009

Hans M. Kristensen and Robert S. Norris, 'Status of World Nuclear Forces', Federation of American Scientists, fas.org/issues/nuclear-weapons/status-world-nuclear-forces

William Lanouette (with Bela Silard), *Genius in the Shadows: A Biography of Leo Szilard*, Scribner, New York, 1992

Egon Larsen, *The Cavendish Laboratory: Nursery of Genius*, Edmund Ward, London, 1962

Rob Linn, *The Spirit of Knowledge: A Social History of the University of Adelaide*, Barr Smith Press, Adelaide, 2011

Malcolm Longair, *Maxwell's Enduring Legacy: A Scientific History of the Cavendish Laboratory*, Cambridge University Press, Cambridge, 2016

Robin McKown, *Giant of the Atom: Ernest Rutherford*, Julian Messner, New York, 1962

Frederick Mann, *Lord Rutherford on the Golf Course*, Cambridge University Press, Cambridge, 1976

Peter Michelmore, *The Swift Years: The Robert Oppenheimer Story*, Dodd, Mead & Co, New York, 1969

Ann Moyal, *Portraits in Science*, National Library of Australia, Canberra, 1994

David Nichols (ed), *Ernie's War: The Best of Ernie Pyle's World War II Dispatches*, Simon & Schuster, New York, 1986

Ted Nield, 'Beyond the Call of Duty', *New Scientist*, Vol 123, No 1681, 9 September 1989, p 77

Robert Norris, *Racing for the Bomb: The True Story of General Leslie R. Groves*, Steerforth Press, South Royalton, Vermont, 2002

Hans C. Ohanian, *Einstein's Mistakes: The Human Failings of Genius*, Norton & Co, New York, 2008

Mark Oliphant, *Rutherford: Recollections of the Cambridge Days*, Elsevier, Amsterdam, 1972

—— 'Some Personal Recollections of Rutherford, the Man', *Notes and Records of the Royal Society*, Vol 27, No 1, August 1972, pp 7–23

—— 'The Beginning: Chadwick and the Neutron', *Bulletin of the Atomic Scientists*, Vol 38, No 10, December 1982, pp 14–18

—— 'Sir Mark Oliphant' in Peter Rolph (ed), *Fifty Years of the Cavity Magnetron: Proceedings of a One-Day Symposium, 21 February 1990*, Birmingham University School of Physics and Space Research, University of Birmingham, Birmingham, 1990

Abraham Pais, *J. Robert Oppenheimer: A Life*, Oxford University Press, New York, 2006

Rudolf Peierls, *Bird of Passage: Recollections of a Physicist*, Princeton University Press, Princeton, 1985

—— *Atomic Histories*, American Institute of Physics, New York, 1997

Jan Polkinghorne, *Mylor: Valley of Dreams*, Lutheran Publishing House, Adelaide, 1991

Diana Preston, *Before the Fall-Out: From Marie Curie to Hiroshima*, Corgi Books, London, 2005

Richard Reeves, *A Force of Nature: The Frontier Genius of Ernest Rutherford*, Atlas & Co, New York, 2008

Richard Rhodes, *The Making of the Atomic Bomb*, Simon & Schuster, New York, 1986

—— *Dark Sun: The Making of the Hydrogen Bomb*, Simon & Schuster, New York, 1995

Anna Rothe (ed), *Current Biography: Who's New and Why*, H.H. Wilson Co, New York, 1951

Albert Rowe, *One Story of Radar*, Cambridge University Press, Cambridge, 1948

Ernest Rutherford, 'Henry Gwyn Jeffreys Moseley', *Nature*, Vol 96, No 2393, 9 September 1915

—— 'Bakerian Lecture: Nuclear Constitution of Atoms', *Proceedings of the Royal Society A*, Vol 97, No 686, 3 June 1920, pp 374–400

—— 'The Electrical Structure of Matter', *Annual Report of the Board of Regents of the Smithsonian Institution 1924*, Publication 2795, Washington, DC, 1925

Silvan Schweber, *Einstein and Oppenheimer: The Meaning of Genius*, Harvard University Press, Cambridge, Massachusetts, 2008

Robert Serber, *Peace and War: Reminiscences of a Life on the Frontiers of Science*, Columbia University Press, New York, 1998

Philippa Mein Smith, *A Concise History of New Zealand*, Cambridge University Press, Cambridge and Melbourne, 2005

C.P. Snow, *The Physicists*, House of Stratus, Looe, Cornwall, 2001

Ferenc Morton Szasz, *British Scientists and the Manhattan Project: The Los Alamos Years*, Macmillan Academic & Professional, Basingstoke, 1992

Thaddeus J. Trenn, *The Self-Splitting Atom: The History of the Rutherford–Soddy Collaboration*, Taylor & Francis, London, 1977

Peter Watson, *A Terrible Beauty: A History of the People and Ideas that Shaped the Modern World*, Weidenfeld & Nicolson, London, 2000

Spencer Weart, 'From the Nuclear Frying Pan into the Global Fire', *Bulletin of the Atomic Scientists*, Vol 48, No 5, June 1992, pp 18–27

Spencer Weart and Gertrud Weiss Szilard, *Leo Szilard: His Version of the Facts*, MIT Press, Cambridge, Massachusetts, 1978

Rajkumari Williamson (ed), *The Making of Physicists*, Adam Hilger, Bristol, 1987

David Wilson, *Rutherford: Simple Genius*, MIT Press, Cambridge, Massachusetts, 1983

ACKNOWLEDGMENTS

As a high school student who found fourth-year physics complex and daunting, I owe much gratitude to the many who helped transmute this project from notion to completion over the course of a couple of years.

From the outset, Cheryl Hoskin, Lee Hayes and Marie Larsen at the University of Adelaide's Rare Books and Special Collections library were unstintingly helpful and endlessly accommodating, even while dealing with the disruption brought by major building renovations.

Professor Malcolm Longair, former head of Cambridge University's Cavendish Laboratory and now its historical curator, has been hugely generous with his time and his insights. His book *Maxwell's Enduring Legacy: A Scientific History of the Cavendish Laboratory* proved an invaluable resource.

Likewise, Dr Robert Whitworth – a retired member of the University of Birmingham's physics staff and overseer of the department's historical collection – willingly shared archival material as well as his knowledge of Edgbaston campus. I am also grateful for the assistance provided by staff at Birmingham's Cadbury Research Library Special Collections and London's National Archives.

In Australia, I am thankful to Monica Oliphant for welcoming me into her home and recounting memories of her esteemed father-in-law. The miracles performed by Scott Forbes and Emma Dowden in editing a manuscript that, in many places, was as impenetrable as a heavy metal nucleus were extraordinary. I also remain indebted to Helen Littleton at HarperCollins for her enthusiasm and encouragement throughout this project, and to my agent, Jacinta di Mase, for her wise counsel and ready support.

Given that researching and writing a book can be a most solitary discipline, I can't express sufficient thanks to my mother, sisters, nephew and niece for providing regular nutritional and emotional sustenance, as well as to my brother-in-law, Ken Lajoie, for intrepidly wading through the manuscript's woolly first draft and kindly claiming that he, rather like those who assessed my school physics exams, could see the destination I was trying to reach even if the means of getting there took some deciphering.

Andrew Ramsey
June 2019

of blast
bag filled full

Time along 46 cm

{45 cm, base 70}

= 1322

Width 60 cm

Discharge

area = ½ · 70 · 45 × 60
        cc
    = 90,000 approx

Area = π
     = π
     = 3

soaked middle
weight of 56 lbs in
min = 240 sec

Veloc =

c.p. second = 90000 / 240
            = ·t·occ per second approx & ab

Since

length between

of small ints
  1·65 cm
½ · 70 · 45 × 60